ust Eyeballed

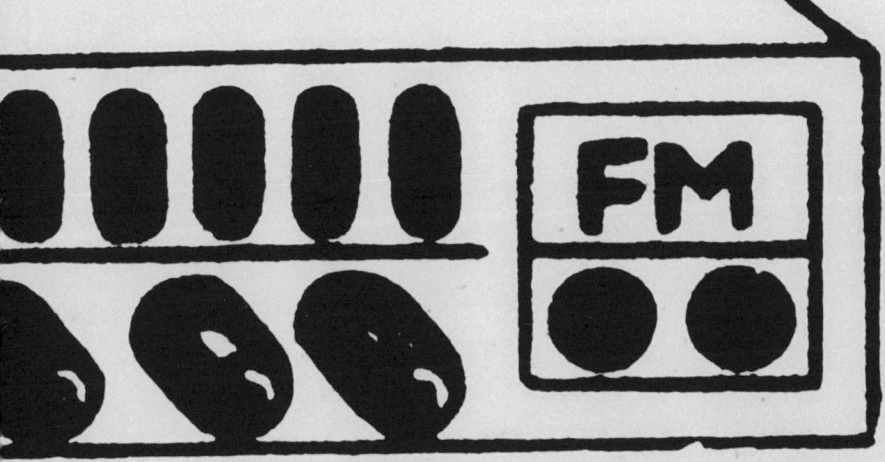

ILLE

Eyeball Cards

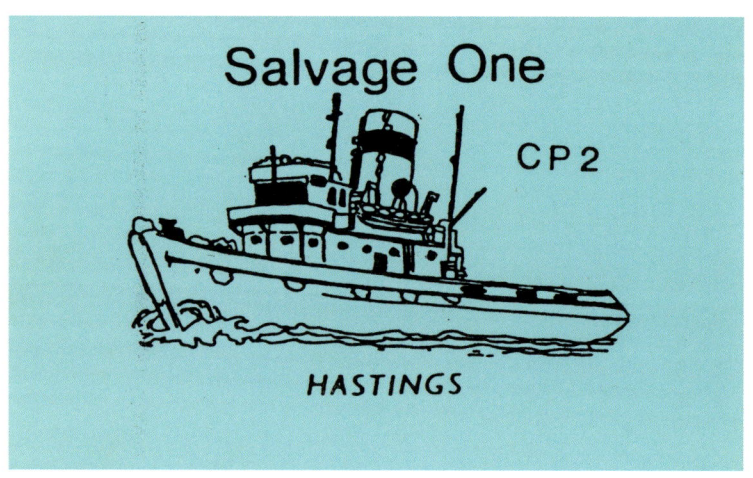

Eyeball Cards
The art of British CB Radio culture

written by William Hogan
with photographs by David Titlow

Four Corners Irregulars
N° I

Appropriately for a device that relied on waves, the UK Citizens Band radio phenomenon came and went like the tide. In the late 1970s and early 1980s, it seemed that every other shed, car and spare room hissed and seethed with the static and chatter coming through these receivers and transmitters. At the time, it was remarkable to think that this magical configuration of metal, solder and crystal could connect you with those nearby, or occasionally, hundreds or even thousands of miles away on dusty roads and distant shores.

Owning a CB rig was open to anyone with £70 and a relaxed approach to poking an elbow over to the wrong side of the law. Pre-legalisation, CB radio existed in a strange legislative hinterland: it was legal to own one, just not to use it. How you got hold of it was relatively easy — there was always a friend of a friend who knew a man who could get a radio and an antenna. In a nutshell, it was the everyman method for short-distance connection in an analogue world.

CB radio created a community for anyone with a rig and a desire to shoot the breeze. Often derided as a geeky or reclusive hobby, for the most part, CB was entirely social, and separate from the more technically-taxing HAM radio scene. The name itself proclaimed it as being for everyone: all citizens, young and old, male and female.

Those who enjoyed it would regularly meet up in person; often in large numbers locally, or just in small groups. The reach of the average set was only a few miles, but, like a spreading vine, each town, and its immediate vicinity had characters who would crossover the conversation into the next area; linking ever-expanding social circles. They were weaving their own communities and creating identities on the airwaves. They were the 'breakers'.

For every breaker, there was a 'handle' — a pseudonym they used to identify themselves on air. Where this verbal identity existed, many took it one stage further and made physical business cards of the CB world: Eyeball cards, the focus of this book. They show the creativity and individuality of the CB-ers themselves and how these qualities found their form — the strange, amusing, sometimes work-a-day or fantastical, or even sinister alter-egos willed into life through these signature cards.

The Eyeball cards in this book have been sourced over years and many miles of the UK. From Essex to East Lothian, from overgrown corners of Wales to the industrial edges of South Shields, we've visited those involved, shot portraits and collected stories by phone, letters, email, social media and CB radios themselves.

The CB radio scene now may only be an echo of what it was, but the artwork, inspiration and social history live on. That's what we wanted to explore and share: how the phenomenon came to be, how it was swept away in a digital swell, and how we should find time to broadcast and celebrate this very British subculture.

William Hogan & David Titlow

EYEBALL! EYEBALL!
YOU HAVE EYEBALLED
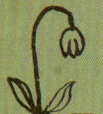
SNOW DROP LEISTON
10-10 TILL I DROP IN AGAIN

EYEBALL! EYEBALL!
— WORKS TOWN BREAKER —
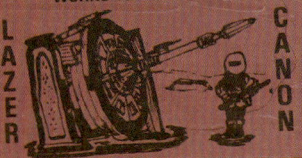
LAZER CANON

ROCK-A-BILLY BABY
BELLS TOWN BREAKER
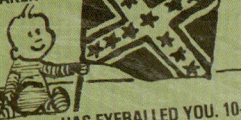
HAS EYEBALLED YOU. 10-10

EYEBALL!   EYEBALL!
ESKIMO
SIZEWELL 20
«10-10 til we do it again!»

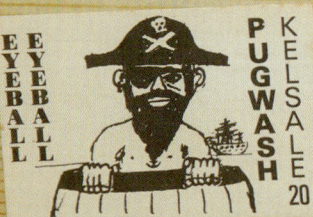
EYEBALL EYEBALL EYEBALL PUGWASH KELSALE 20

EYEBALL! EYEBALL
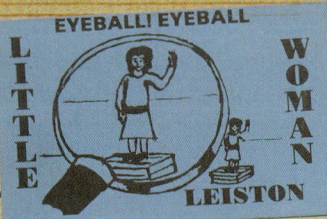
LITTLE WOMAN LEISTON

EYEBALL! EYEBALL!
YOU HAVE EYEBALLED
MIDGET-MAN
OF LEISTON 20
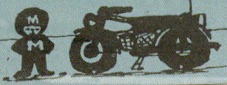
10-10 TILL WE BREAK AGAIN

C·B
EYEBALL! EYEBALL!
GRASS-HOPPER
LEISTON
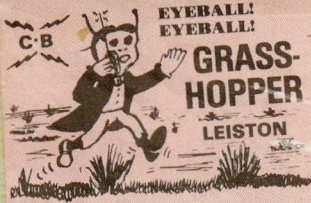

BASKET MAKER
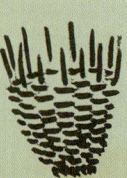
EYEBALLED

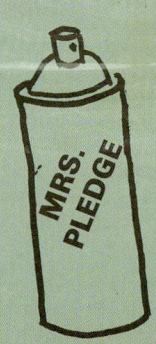
MRS. PLEDGE
EYEBALLED

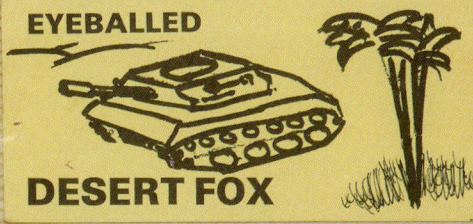

EYEBALLED

DESERT FOX

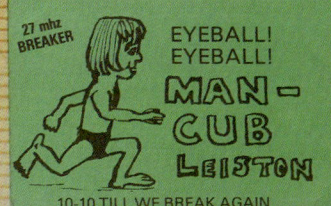

27 mhz BREAKER

EYEBALL! EYEBALL!

MAN-CUB LEISTON

10-10 TILL WE BREAK AGAIN

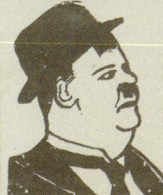

EYEBALL! EYEBALL!

YOU HAVE JUST EYEBALLED

HARDY-MAN of LEISTON

10-10 TILL WE BREAK AGAIN

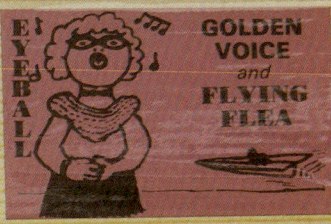

EYEBALL

GOLDEN VOICE and FLYING FLEA

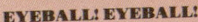

CONGRATULATIONS!
YOU HAVE JUST EYEBALLED

BEANY

Works Town Breaker

10-10 TILL WE BREAK AGAIN

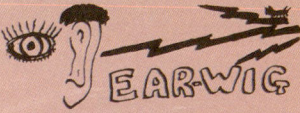

EYEBALL! EYEBALL!

EARWIG

—THORPENESS BREAKER—

PING-PONG — WE'RE GONE

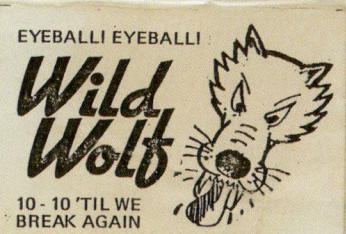

EYEBALL! EYEBALL!

Wild Wolf

10-10 'TIL WE BREAK AGAIN

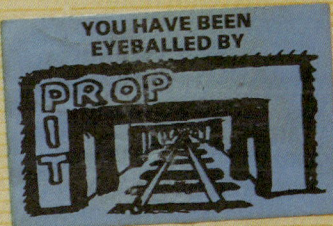

YOU HAVE BEEN EYEBALLED BY

PROP PIT

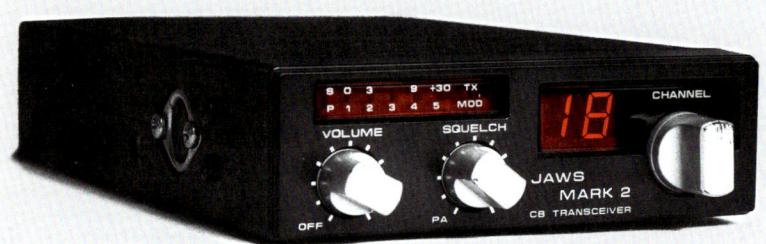

From A Ripple To A Wavelength   William Hogan

With its whiff of officialdom, association with government bodies and utilitarianism, up until the mid-1940s, the reputation of two-way radios was as flexible as a starched collar. The rigs themselves were cumbersome yet internally delicate valve-operated hunks of metal found only within civic offices, logistical transport, or military facilities. The matter-of-fact transfer of information was always at the heart of the medium, and mainly reserved for those doing some type of buttoned-up public service.

So it wasn't until Al Gross devised and patented the portable 'personal radio' in America in 1938 — that later became known as the Walkie-Talkie — that the paradigm began to switch. In the late 1940s, Gross formed the Citizens Radio Corporation, which set to work manufacturing two-way radios for the general public's personal use. The Federal Communications Commission (FCC) granted permission for the use to the public on a less-prevalent radio wavelength, and the new Citizen's Band radio was born.

The groundswell for CB in the USA, was shaped by geography as much as economic or practical reasons. Vast stretches of open road snaking across the middle of America meant the medium made sense between long-distance truckers hungry for conversation to punctuate long journeys.

What tipped it from a nice-to-have device to being the standard piece of trucker kit was the oil crisis in 1973. Fuel shortages and rationing struck every part of the nation. As a counter-measure, the US government decreed a 55mph speed limit. For truck drivers, every incremental delay ate away at their livelihood. With the resulting fuel price hike, the nation's infrastructure began to lose momentum. So truckers took matters into their own hands by embracing CB in vast numbers — their exchanges made sure they were first to get to laden fuel pumps ahead of other gas-guzzling vehicles. Importantly, it allowed them to strike in huge swathes in complete coordination.

A Jaws Mark 2 from the late 70s/early 80s. A classic, reliable 40 channel AM CB Radio.

What emerged was a tangible sense of brotherhood in the trucker community. For those who'd cottoned on and bought one it was a rebellion against 'The Man'. An outlaw attitude had found its amplifier, and CB became the true voice for the people's disdain.

Word spread globally, but the real bump up to national and by default, global awareness came from an unlikely source: Early 1976 saw First Lady Betty Ford granted a temporary Citizen's Band radio license and she began campaigning over the airwaves.

While there were no restrictions on political discourse on CB radio, some users took exception. This fuelled a national debate in the USA, and pushed the public profile of CB radio even higher. This was helped further when she commissioned a QSL card from a popular CB artist known as 'Brushstroke'. These CB calling cards, which were the precursor for the UK's Eyeball cards and a format favoured by HAM radio enthusiasts, also experienced a major boost.

Sent between breakers who'd made a connection over the airwaves, QSL cards differed through the level of detail they would usually include, such as the breaker's handle or call sign, the station registered number,

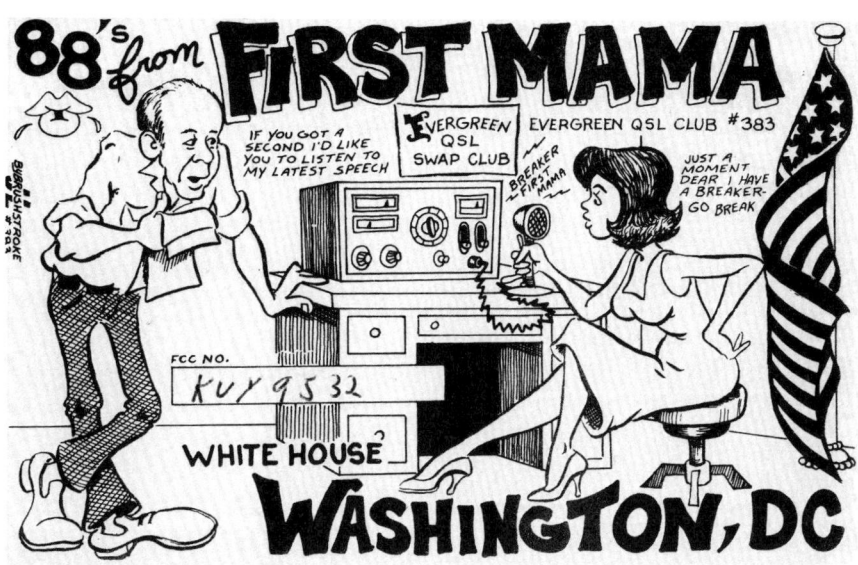

Betty Ford's QSL card.

radio frequency or band used and a signal report. They span the globe and are another universe of postcard-sized visual flights of fancy in their design. They generally made their way around the world by post. In technical terms, they are written confirmation that a two-way radio communication between amateur radio 'stations' (HAM radio in other words) has taken place. ('QSL' is radio shorthand for 'transmission received').

By the time Eyeball cards came to be established in Britain — in their own irreverent style — QSL art had been around for a long time and laid the groundwork for abstract handles and arch graphic representations.

Eyeball cards are strictly a British invention, to be exchanged when two CB-ers actually met. Whereas the QSL card would be mailed globally, Eyeball cards were a visual and entirely localised confirmation, the real-life 'eyeball'.

With awareness of CB building in the US at pace from the mid 1970s, it was only a matter of time before the associated outlaw connotations were picked up in Britain. A very influential TV show in CB terms, *The Dukes of Hazzard* aired stateside to over 21 million regular viewers, and Britain broadcast the show a few months after its US release.

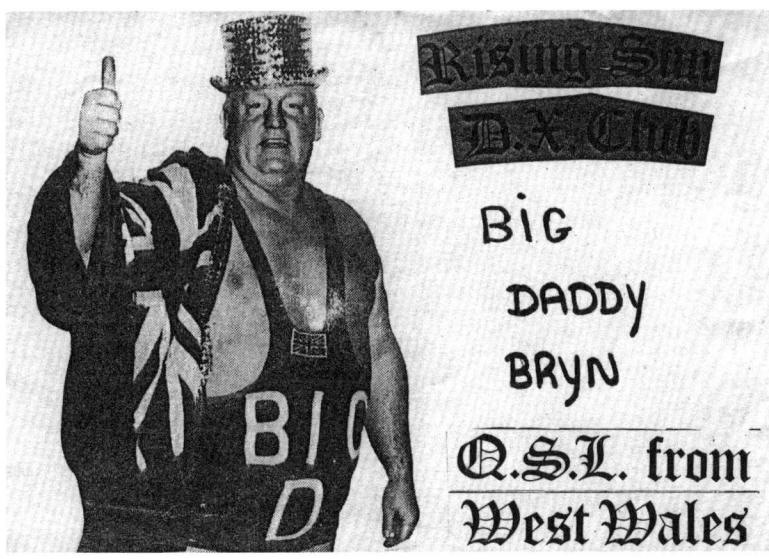

A British QSL card. These were the precursors to the Eyeball card.

The show was an instant hit, with CBs taking centre stage and as a key tool in the loveable rebellion of the lead characters, Bo and Duke, and their high-octane high jinks. Coupled with antagonists Boss Hog and Sheriff Rosco Coltrane's exasperated exchanges from scuttled police cars showed that CB radios were potentially involved in the mechanics of a more exciting rural life.

Contemporary British CB-enthusiast magazines depicted a glamourous hinterland of women in tight t-shirts and cut-off denim – the eponymous Daisy Dukes. Showing that the glossy aspirations ran as high up the leg as the shorts — the transmission of an American Dream was finding ambitious UK form.

Under Margaret Thatcher, there were major fissures in the economy, race relations and internationally — all seemingly handled with the Iron Lady's characteristic iron fists. The spectre of nuclear destruction also loomed large with the release of the public information film *Protect & Survive* in 1980.

Arguably, such a form of escapism in the shape of CB radio was just the tonic Britain needed.

For some, apocalyptic fears and their inner survivalist came to the surface to pique an interest in personal radio communications. For 'preppers', CB radio is an essential piece of kit — being the only form of technology that would likely work in a 'world-ending' event with the collapse of satellite and digital networks. Anecdotally, some current British CB-ers also reference the hardy analogue radio wave as one form of communication salvation in a disaster scenario.

※

Liverpool was traditionally the conduit for Americana finding its way to British shores in all its forms since the 1950s. Alongside records, clothing and attitudes, Liverpool docks became a major unloading point for the new sub culture of CB radios in the UK.

Painfully for the flailing domestic economy, British manufacturers missed out on a potentially lucrative revenue stream because of the legislative problems with actually using them. Filling the demand vacuum, far-east countries like Taiwan and Japan eventually stole market share and dominance against a previously prolific US industry.

There was, however, plenty of money to be made in sales. One former illegal seller with the handle 'Cowboy' recalled: "It was heavy amounts of

A CB magazine from North Wales, 1982.

ON THE AIR IN NORTH WALES

60p

NEW CB MAG

With Black INK

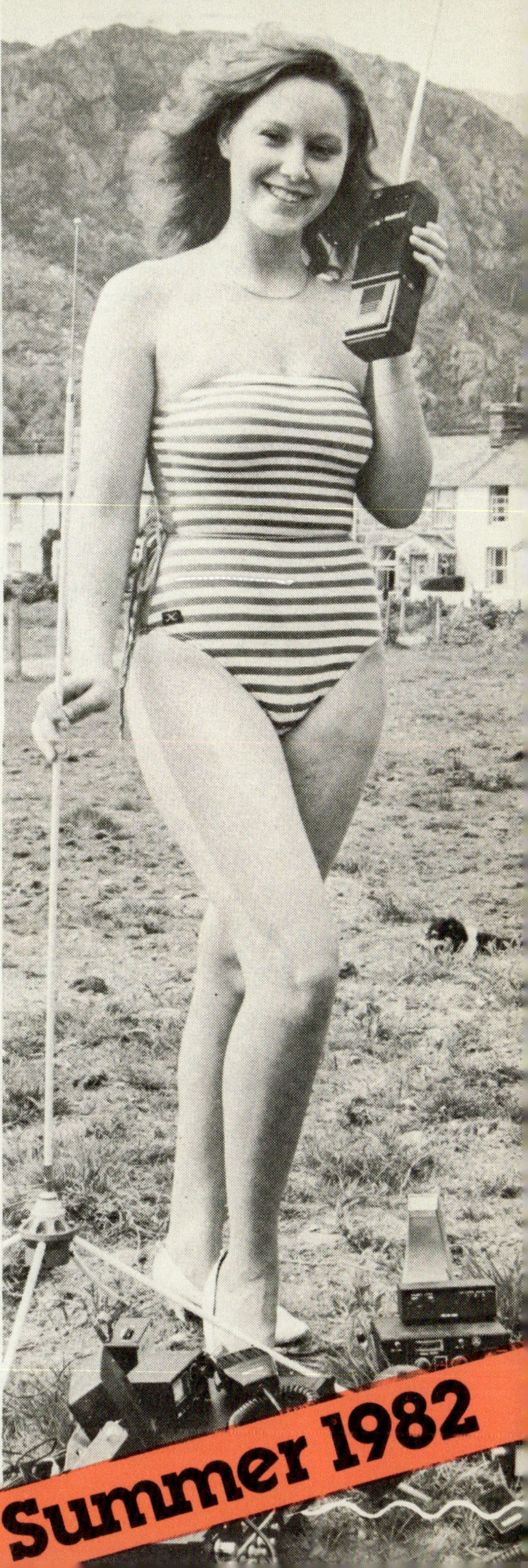

Summer 1982

money changing hands. Probably running to a couple of hundreds of thousands each week." The consignments of radio equipment were then transported throughout distribution contacts nationwide. Not that Liverpool docks had the monopoly on CBs: many major ports saw hundreds upon thousands coming through every week.

The news program *Reporting London* broadcast a CB special in 1981 detailing the popularity surge. One shop owner, who fitted and sold CB equipment and paraphernalia, recounts turning round 10,000 aerials a week, and £250,000 worth of business every month — a 70/30 mix of illegal to legal sales.

The figure of the CB wheeler-dealer was alluring. In Colchester, Simon Morgan aka 'Pieman' was on day release (a national initiative to help the vast numbers of unemployed young people into work) at a local garage in 1981 when he first came into contact with CB rigs. He recalls asking them how to get one, and that 'anything CB was available through a "friend of a friend". Rigs like Superstars and the Cobras were significant pieces of kit, so how hundreds were readily available was anyone's guess.' CB radio shops — usually under the banner of radio repair businesses — sprang up nationally. Simon mentions that there were three in Colchester alone. 'The guy who ran one shop, Peter Hopkins, had the handle "Magnum". Another on Barrack street, went by the handle "Wallet" but that may have been to do with how well he was doing out of them.'

The rigs themselves sounded as though they were straight out of a Western. Manufacturers gave them titles such as the 'Cobra' or the 'Colt' — evocative of spit and leather from the Midwest. It was reflective of the tough-guy image that suited the rebellious heroes that Britain was clearly holding out for.

Despite the male-centric marketing, CB radio had a huge female following. Love matches and eventually marriages began on the airwaves and it wasn't uncommon to hear many female breakers chatting and holding their own in a predominantly male universe. And the number of female Eyeball cards showed that women were very much part of its cultural fabric.

On TV news in mid-1980, one breaker gleefully (and perhaps somewhat optimistically) estimated that there were around two million users in London alone. The Department of Trade and Industry (DTI) whose job it was to oversee crackdowns on illegal CB operation, records that only around 9,000 sets were confiscated that year. The police, by all anecdotal accounts, had bigger fish to fry and so left the clampdown to 'Busbies' — big listening aerials on detector or 'Porcupine' vans that would twist around to point to where the signal was coming from in almost cartoon style. Run by the Post Office, there were hefty fines of around £400 and equipment would be confiscated if detected.

Far from putting CB-ers off, the whiff of illegality meant the allure was stronger than ever. It was a game of cat and mouse, intrigue and evasion, and instances where chases occurred were not uncommon. 'Pieman' recalls an incident where he was broadcasting from his van (fitted with an eight-foot 'whip' or aerial) and was approached by two DTI inspectors for a 'bust'. With echoes of a chase scene from *The Dukes of Hazzard*, the inspectors gave an unsuccessful chase in their gold Granada.

There were tales of ingenious mechanisms to hoist and lower aerials. Suffolk breaker Jenny aka 'Fruitcake' remembers uncoupling the 'boots' (signal boosters) if they suspected detection. The scaffold-pole construction built by her late husband Bob aka 'Sugarbeet' meant evidence of CB use could quickly be minimised.

The main source of the backlash was from people claiming their radio and TV reception was affected — occasionally with entire conversations being heard through the sets themselves, sometimes from as far afield as Holland. Even when it was turned off, stereo equipment wasn't always safe from the chatter on the airwaves. Breakers using a tweaked wattage that exceeded the original factory settings would be stigmatised because of incidents like these.

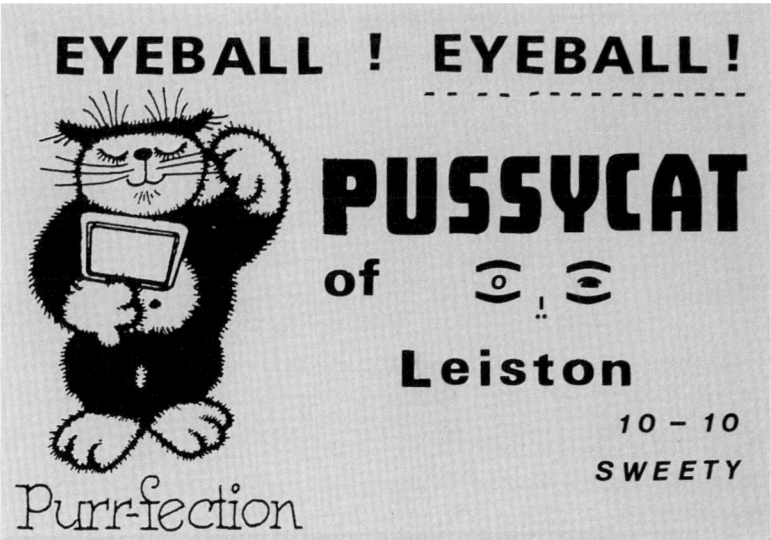

Even in its 'Wild West' state, CB radio was self-regulating. There were and still are some hard-and-fast rules: it's never to be used for entertainment purposes, such as playing music, and no swearing or false information can be shared. Breakers were enforcing these rules themselves, ignoring requests to talk by those frequently contravening the unwritten rules and remonstrating with those not adhering to them.

Topics of conversation ranged from what rig people were operating on, their location (coded, naturally) local gossip, practical travel information, weather, meet ups, news of busts and anything they found amusing — playful chatter in other words. Talking further afield than say 20 miles meant finding higher ground for better reception, so mobile CBs fitted to vehicles became a natural next step. There were modifications to cars, vans and mopeds and even bicycles. Mark Atkinson aka 'Bandit' recalls affixing an aerial on his Chopper pushbike, with burglar alarm batteries nestling on his bicycle rack and the CB radio on the handlebars. He would then ride to the top of the tallest mountain he could find and often talk to breakers hundreds of miles away.

Eyeball cards and QSL cards were at the heart of the CB community and galvanised on-air popularity. Martin Cook, another breaker with the handle 'Bandit', recalls his local club with around 200+ members. The 'Cheltenham & District Breakers Association' (CBDA) held monthly meetings and various ad-hoc meets to swap cards and introduce new members. 'Part of the club dues was put into a "Bust Fund". Anyone who got caught was guaranteed to be back on the air within 24 hours. We also paid their fines.' Other clubs, such as Barnsley's Breaker Club recalled by the breaker known as 'Fidget', cites just over 1000 members at its height, with activities like 'Foxhunting': a hide-and-seek game using CB-equipped vehicles that involved triangulating the position of the 'fox' who would broadcast sporadically and the 'hunters' would use their signal meters to determine where they were hiding. At one point, national competitions were organised with competition prize money offered for the fastest.

While the airwaves buzzed with chatter and relationships were created, there was still an undercurrent of paranoia in divulging too many details about the users — especially locations. This elusive nature was fun in tight-knit rural communities where it was often common knowledge who was behind the handles, which then crossed over to the pub, or the local supermarkets. The principle behind the 40 open channels defines that anyone could 'earwig', so silent audio voyeurs were a consideration too. Newcastle breaker 'Dandelion' recalls her CB identity becoming her real-life handle: 'The local bobby was standing behind me in the queue for the cinema and said "Afternoon, Dandelion." Of course, I blushed, but we had a good laugh about it.' While countryside communities across the UK (anecdotally) took a more relaxed approach to enforcing the law, some more built-up areas took the fines and confiscation of equipment seriously. 'Black Hat' recalls his Birmingham flat being raided and his equipment being taken indefinitely. 'There didn't seem to be any hard-and-fast rules about what happened when the DTI decided to intervene.' With so many breakers operating, it was a numbers game for inspectors — a lottery based on who they happened to hear when driving by in their less-than-stealthy vans, and possibly 'a tip-off by disgruntled neighbours who may have heard radio chatter late at night' or, he playfully concedes, 'my wife could have called in the bust as my attention was constantly taken up by my CB activity.' Mark Alden aka 'Praying Mantis' mentions: 'Using the CB illegally was probably the same kind of feeling as not paying your TV license: that any minute someone is gonna knock on your front door.'

In many ways, CB radio had become a victim of its own popularity — the airwaves were jam-packed, and it was often difficult to hold a conversation with the sheer numbers of breakers on all available channels. Pressure was mounting for legislation. In 1979, around sixty countries had legal CB radio, and some of those were behind the Iron Curtain. The same year, the Citizen's Band Association (CBA) was gathering influence, applying pressure to the Home Secretary to allocate a frequency.

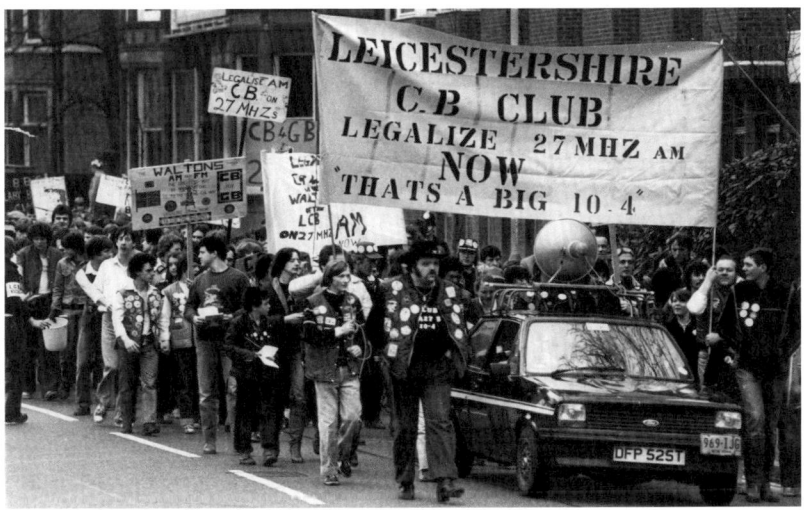

Leicestershire CB enthusiasts march in 1981.

Over the course of 1979 and 1980, thousands marched to legalise CB Radio across the UK. In 1980, demonstrations stretching as broadly as Donegal, Ireland, to Brighton, Manchester, London, and everywhere in between saw trails of people and vehicles take to the streets.

In 1981, 27 FM was chosen as the legal channel for CB radio. The AM frequency had been used up until now by the majority of CB users, and as soon as legality was passed, the channels were swamped. The popularity was reflected in the best-selling Christmas presents for 1981 — seeing CB radios feature high in the ranking. In Ireland alone, RTE stated in a 1981 news report that £2 million a year was being spent on legal CB radios. With the influx of people on the frequency, breakers resorted back to AM to find space to chat, and in all probability, an echo of a more illicit and exciting time, pre-legalislation.

Ironically, once CB was legalised, and after an initial surge of activity, users' interest dropped from a great height — the gamble with the law had vanished along with the allure and the image built on rebellion.

Within a year or so, the activity was in sharp decline with more second-hand rigs on sale than ever before. Fast forward to the advent of cordless phones (ironically, another invention by CB radio creator Al Gross) and the subsequent appetite for mobile phones and the digital march and CB culture — as it once was — is essentially obsolete.

Barring some disparate and small CB communities operating sporadically and spread very far apart, what now remains from what was once a personal communication phenomenon are the Eyeball cards.

CB users' handles were drawn, pressed, Xeroxed and swapped at 'meets', larger group events, or shared in person with new friends made on the airwaves. Even playgrounds seemed to echo with the immortalised identities of local breakers — taking on Panini stickers as collectibles to be swapped, bartered and liberated from living rooms and garden sheds nationwide.

The fact that CB in Britain began illegally meant that these calling cards took on a gravity all of their own. As a youngster growing up at the time, it seemed that this world materialised through half-whispers from older siblings, and dads surreptitiously fixing aerials to outhouses and Ford Cortinas; all done with a nod and a knowing wink.

With the philosophy of an in-joke against the establishment and a straight-faced gaze into the unknown — the world of CB radios and Eyeball cards was a Klondike of faded-denim dreams, and dusty American roads dreamt from forgotten corners of Britain. This wasn't just a phenomenon

You Have Just Eyeballed

GOALIE

73s   88s

Willesborough Twenty

Take Care
Till We Meet Again

imported piecemeal from the United States; this was taken into British homes and hearts wholesale. The precedent set by the outlandish QSL cards of rural states, and the breaker's language and style found more than its translation this side of the pond — it discovered its own identity altogether in the myriad designs.

The cards themselves became the definition of accessibility to a coded world — rich with cartoon iconography and a DIY cut and paste ethos that inspired hundreds of thousands of people to take part.

Often, the cards would give clues as to the owner's job. For some, it would represent their hobby. For others an insight into their fantasies. To the uninitiated, they appeared as nonsense, or worse, have a whiff of Amateur Photographer about them: all bums and boobs and seaside sex.

Like coy swingers strolling down a suburban drive, the pseudonyms of Eyeball cards make for an eclectic parade. The unassuming 'Little Bo Peep', and the evocative 'Moonlight Gambler & Devil Woman' rub shoulders with 'Randy Andy', 'Black Bomber & Bombshell' and 'Blue Eyes'. It's a melange of ideologies, of working titles and hopes that try to break free from geography, despite — or because of — area names like 'Container Town' and 'Works Town'.

'Pieman' aka Simon Morgan recalls visiting a printer doubling as an Eyeball card designer in the back streets of Colchester. 'His desk and floor area were covered by hundreds upon hundreds of stacks of cards that people had ordered. It was him doing, probably, the whole area.' Across Britain at the time, another breaker with the handle 'Redneck' recalls it being around a fiver for a couple of hundred cards. He recalls: 'We didn't resent paying for them, no. Everyone had one, so it was just part of it, and no one would turn up to a weekly meet without a wad of them in their back pocket." Jenny Dye, aka 'Fruitcake', recalls the local wheeler-dealer called Brian Hales selling CB rigs and printing Eyeball cards in the back of his general store in Leiston, Suffolk. 'He did a roaring trade,' she recounts.

There are so many reasons why a breaker's name came to be chosen — from the obvious to the tangential. Fruitcake's late husband Bob was 'Sugar Beet' from the agricultural connection — himself a seasonal sugar beet farmer. Hers became Fruitcake, 'not for my baking abilities, but because one of my offspring suggested I was nutty as one.' Sons Matt and Simon were (respectively) 'Eskimo' — after building a particularly impressive igloo with a friend in winter, and 'Toecutter' — after a character in the film Mad Max.

While there were no set formats for Eyeball cards — size variations, portrait, landscape, textured edges, playing-card finish, matt, gloss, card stock, images, words, or both — different areas developed their own style of Eyeball cards. It seemed there was a handful of local artists in some cases taking on commissions, or breakers going heavier with the DIY feel, using pen, pencil, typewriters — indeed anything that would mark out their individuality best with the materials at hand.

Just as the paranoia of getting caught influenced how candid breakers could be on air, the cards' details also made the world more esoteric. 'The last thing you wanted to do was give away your '20' (breakers' Q Code for location),' cites 'Hummingbird' a Carlisle breaker. Aptly for his handle, the 'Budgie Perch' was code for the Pennines where he would occasionally broadcast to get the most range. 'Obviously, the lengths we went to avoid detection on air meant your Eyeball card shouldn't give the game away.'

There were myriad nicknames for places: Edinburgh was 'Castle City', Lincoln was 'Cathedral City', Cornwall was 'Surf County', Kent was 'Hop County' etc. Larger towns even assumed their own identity: Dyfed in Wales was 'Sausage Town', Bexhill was 'Foggy Town', Hastings was 'Ghost Town', Diss in Norfolk was 'Dodge City' etc. Within these areas, even more localised places took on their own handles. In Suffolk, Eyeball cards feature a town-by-town rundown with nicknames like 'Antiques Town', 'Gravel Town', 'Saxon Town', 'Donkey Town', 'Bells Town' etc., all marked by industrial or historical associations. The towns' and villages' handles demonstrated how esoteric and intriguing the CB world had become.

YOU   HAVE   EYEBALLED

Coral
&
Concho

SOMPTING 20

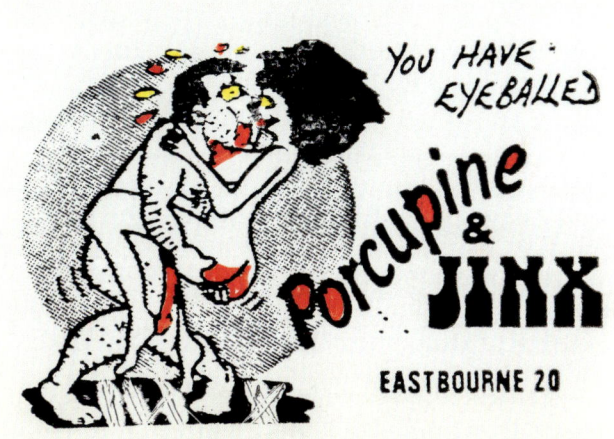

Lollipop

&

Phantom Swinger

Home 20   Ulcombe

Breakers were meeting up after chatting on the air and relationships were forming — encapsulated by couples sharing the same Eyeball card. Who decided the design in the pairing is open to interpretation, but from the olde-worlde font of 'Coral & Concho', Eastbourne's in-flagrante 'Porcupine and Jinx', the questionable monikers of 'Lollipop & Phantom Swinger' to the minimalist and elusive 'Blue Fox & Silver Cloud', it's an individualised art that's moved into the realm of couple's combined fantasy identity.

What's apparent in the wording in particular is the affection breakers have for the cards and each breaker. '3s and 8s' (which means 'love and kisses') and 'Take care because we care' are frequently found across the collection of cards. The immortal ' 10-10' — again taken from the break-ers' lexicon, the Q Code — means the CB user will not continue broadcast-ing but keep listening in — invariably followed by the half rhyme ' 10-10 'til we meet again'. As a collection, you can almost hear the voices from the cards, appearing from the hissing ether like half-remembered punchlines, or a playful wink to another, more exciting life for the owners. As digital dominates attention spans and lives of young and old, whatever happens to analogue's CB radio will live on forever through Eyeball cards — an art form that will never 'be down, be gone'.

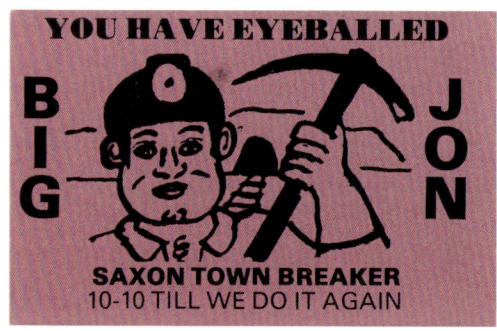

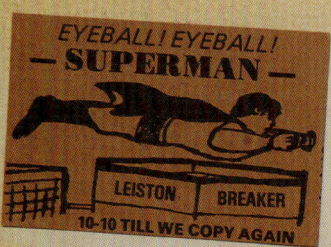

EYEBALL! EYEBALL!
— SUPERMAN —
LEISTON BREAKER
10-10 TILL WE COPY AGAIN

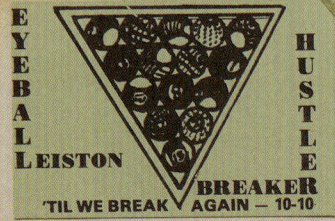

EYEBALL LEISTON HUSTLE BREAKER
'TIL WE BREAK AGAIN — 10-10

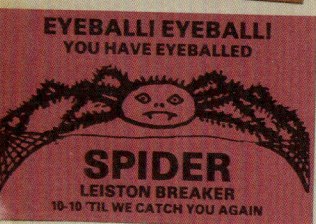

EYEBALL! EYEBALL!
YOU HAVE EYEBALLED
SPIDER
LEISTON BREAKER
10-10 'TIL WE CATCH YOU AGAIN

Eyeball   Eyeball
YOU HAVE JUST EYEBALLED
DART PLAYER
Works Town Breaker
10-10 TILL WE BREAK AGAIN

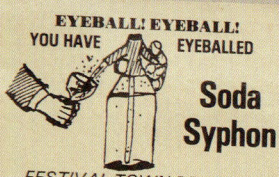

EYEBALL! EYEBALL!
YOU HAVE   EYEBALLED
Soda Syphon
FESTIVAL TOWN BREAKER
10-10 'TIL WE SPLASH AGAIN

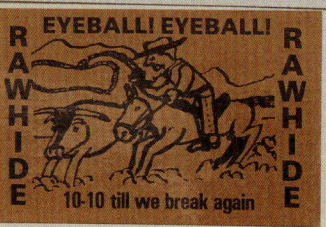

EYEBALL! EYEBALL!
RAWHIDE   RAWHIDE
10-10 till we break again

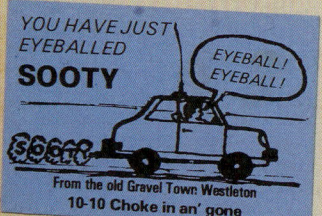

YOU HAVE JUST EYEBALLED
SOOTY
EYEBALL! EYEBALL!
From the old Gravel Town Westleton
10-10 Choke in an' gone

CONGRATULATIONS
YOU'VE EYEBALLED
NINE CARAT
ALDEBURGH BREAKER

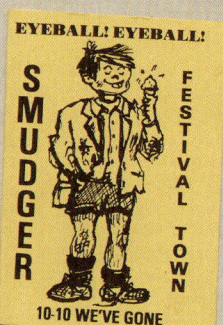
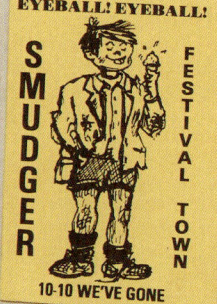

EYEBALL! EYEBALL!
SMUDGER   FESTIVAL TOWN
10-10 WE'VE GONE

EYEBALL!
BLACKSHADOW   CHESHUNT 20
10-10 COPY AGAIN

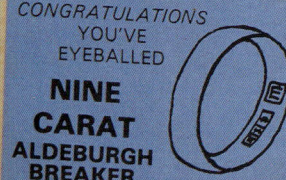

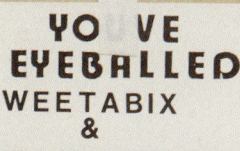

YOU'VE
EYEBALLED
WEETABIX
&
GRAPEFRUIT
10~10

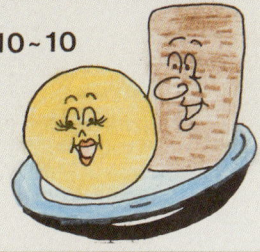

EYEBALL! EYEBALL!
You Have Just Eyeballed
Diddy David
10-10 Till We Break Again

Eyeball
IRONMONGER
OF
WALBERSWICK
10-10 GOOD BUDDY

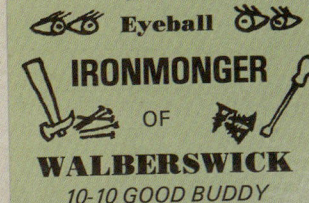

EYEBALLED
POACHER
CATCHER
10·10

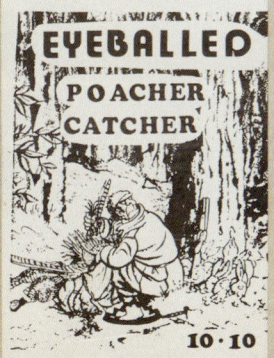

CONGRATULATIONS!
YOU HAVE JUST EYEBALLED
FIVE TONE
Works Town Breaker
10-10 TILL WE BREAK AGAIN

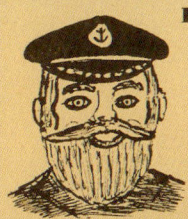

EYEBALL! BRIGHTEYES
(CASTLETOWN)

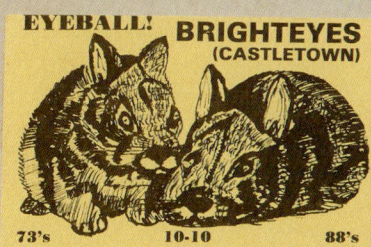

73's   10-10   88's

EYEBALL!
EYEBALL!
YOU HAVE JUST
EYEBALLED
Captain
Birdseye
ALDEBURGH
10-10 'TIL WE
BREAK AGAIN

CONGRATULATIONS!
YOU HAVE JUST EYEBALLED
BIG DADDY
Maidstone Kent Breaker
10-10 TILL WE BREAK AGAIN

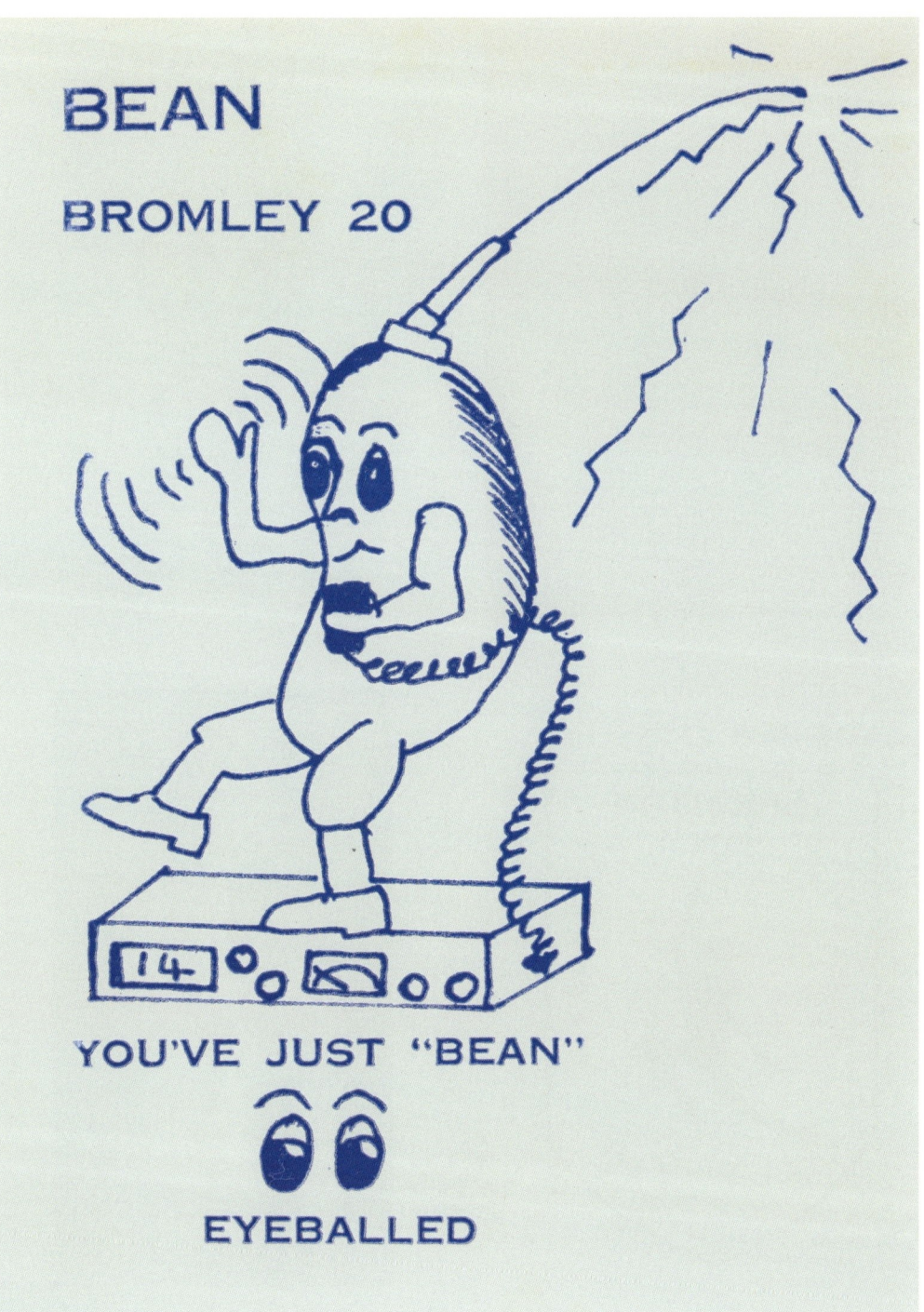

BEAN

BROMLEY 20

YOU'VE JUST "BEAN"

EYEBALLED

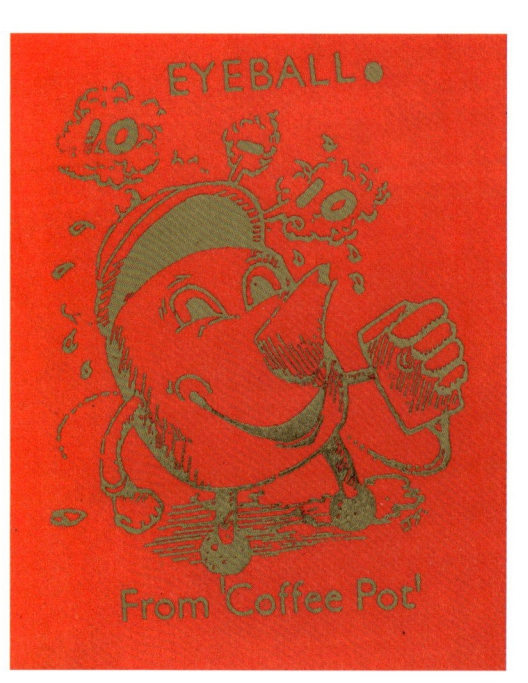

EYEBALL

From 'Coffee Pot'

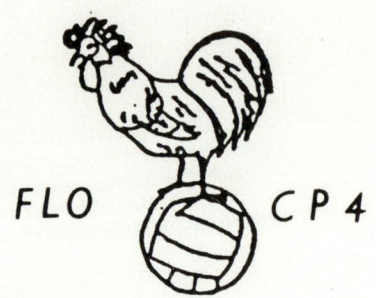

Hotspur

FLO   CP 4

HASTINGS   20

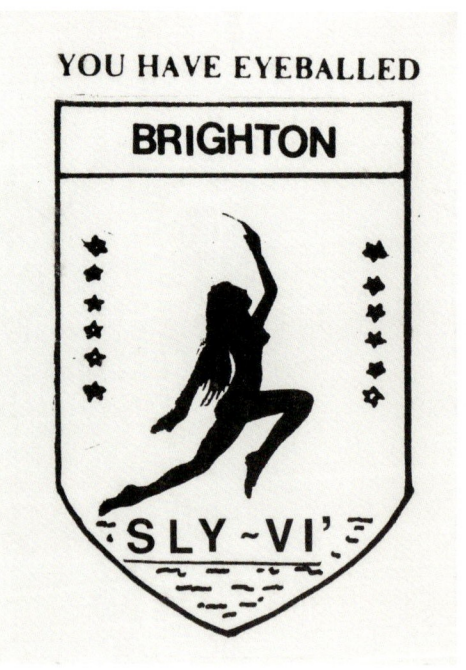

YOU HAVE EYEBALLED

BRIGHTON

'SLY ~ VI'

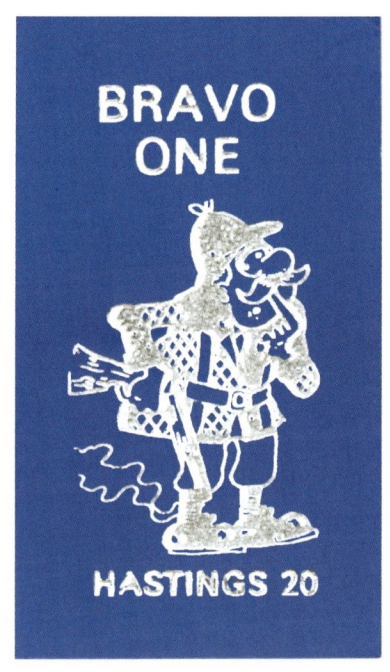

BRAVO ONE

HASTINGS 20

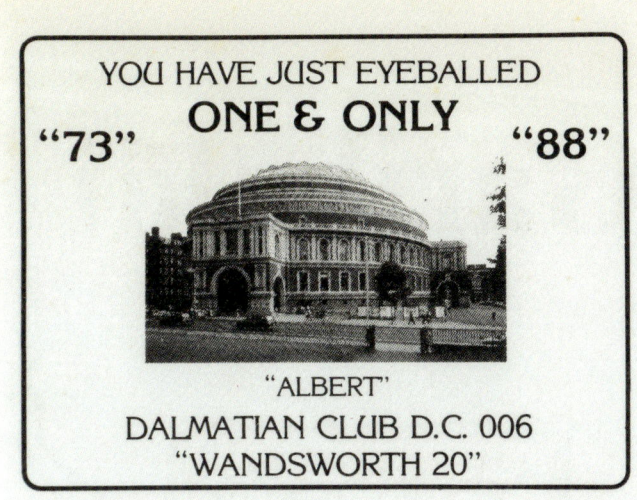

YOU HAVE JUST EYEBALLED
ONE & ONLY
"73"   "88"

"ALBERT"
DALMATIAN CLUB D.C. 006
"WANDSWORTH 20"

EYEBALL!   EYEBALL!
YOU HAVE JUST EYEBALLED

PEPSI

CASTLE TOWN 25 CLUB
10-10 Till we break again

You have
JUST
EYEBALL
PET
LOVER

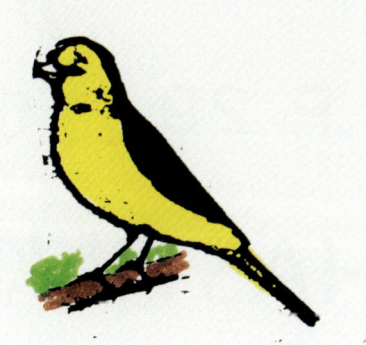

WIMBLEDON 20.

Garibaldi

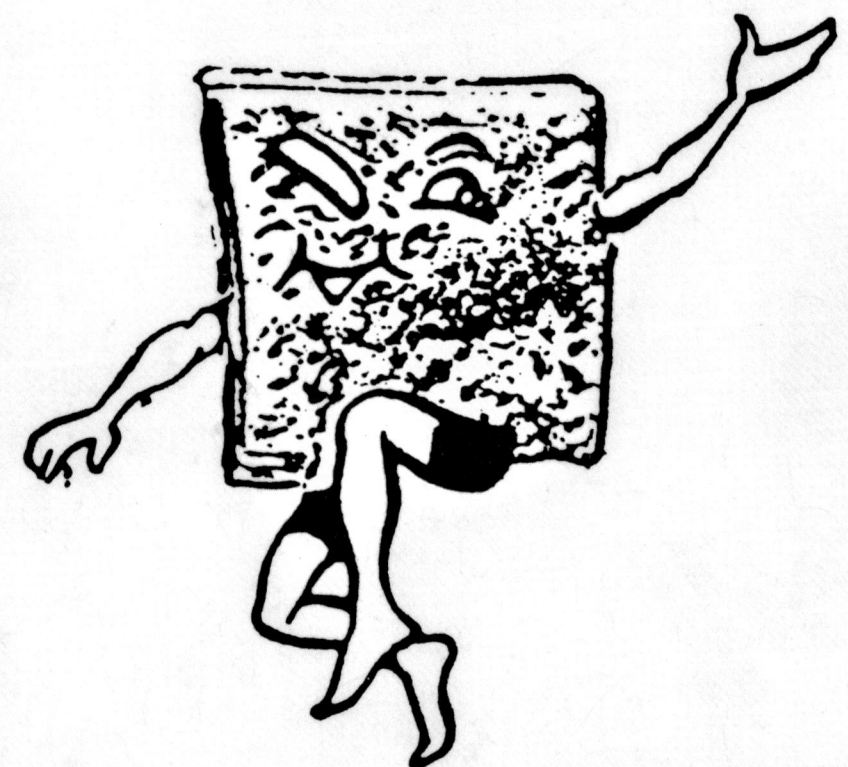

C P 29   CHARLES

HASTINGS 20

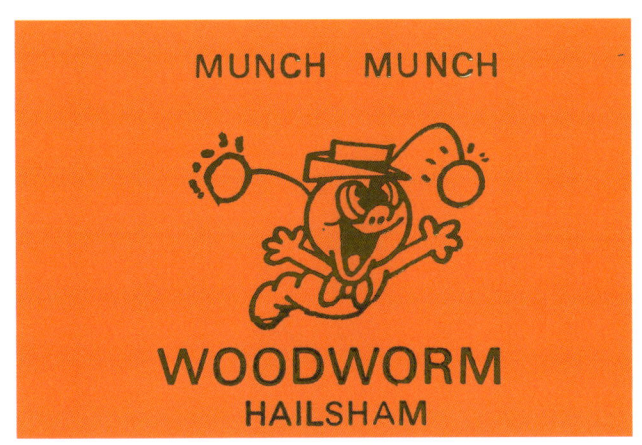

MUNCH MUNCH

WOODWORM

HAILSHAM

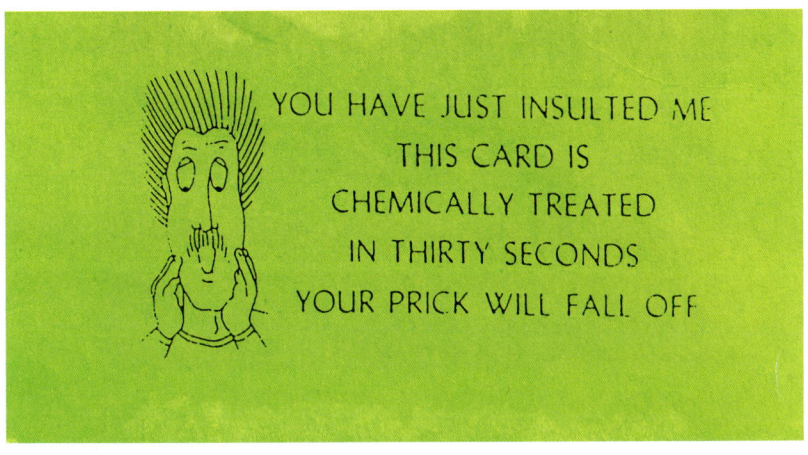

YOU HAVE JUST INSULTED ME
THIS CARD IS
CHEMICALLY TREATED
IN THIRTY SECONDS
YOUR PRICK WILL FALL OFF

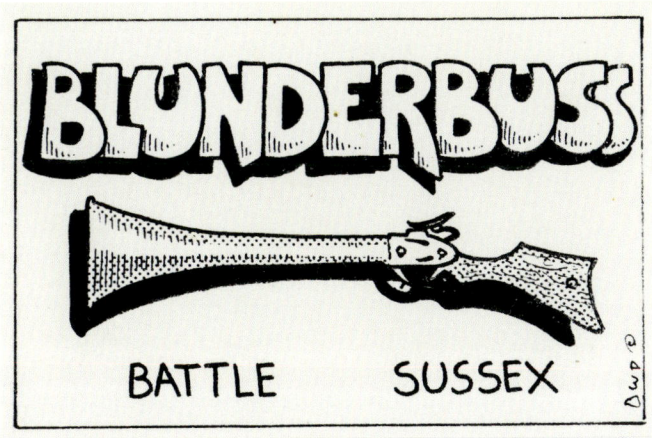

BLUNDERBUSS

BATTLE   SUSSEX

YOU HAVE JUST EYEBALLED

LEN. The Pedlar

ECHO KILO 26 HAWKHURST 20

10 - 10 TILL WE DO IT AGAIN

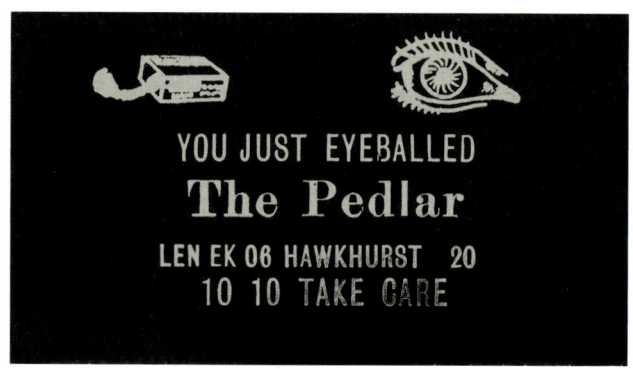

YOU JUST EYEBALLED
The Pedlar
LEN EK 06 HAWKHURST   20
10 10 TAKE CARE

YOU HAVE JUST EYEBALLED

EUROPA 1
(Silvia)

FROM THE REDHILL 20

10-10   TILL WE MEET AGAIN

SIGNALMAN

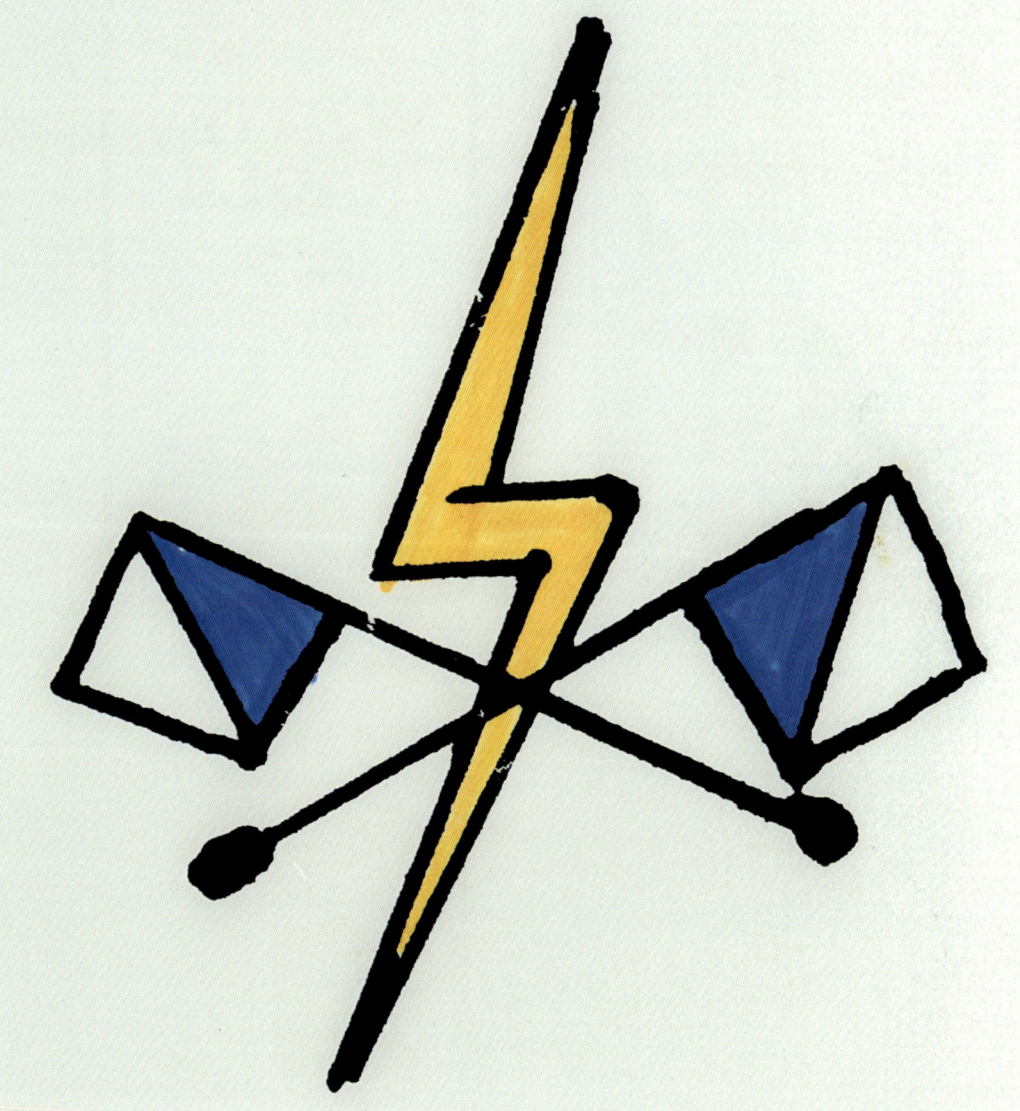

Hawkhurst (Ken)

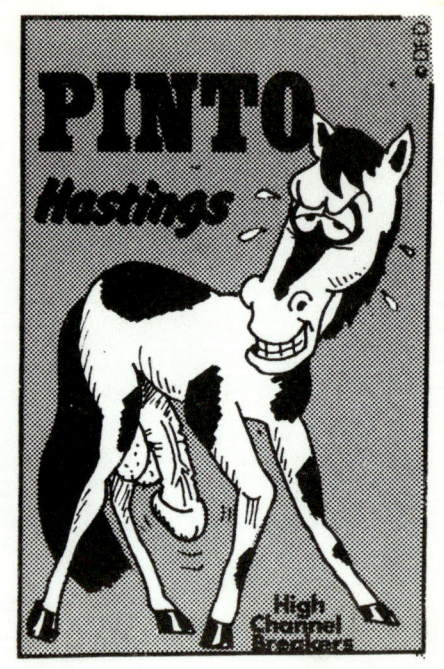

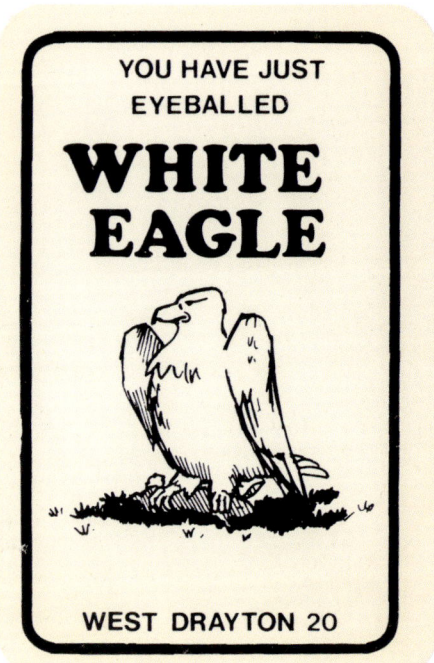

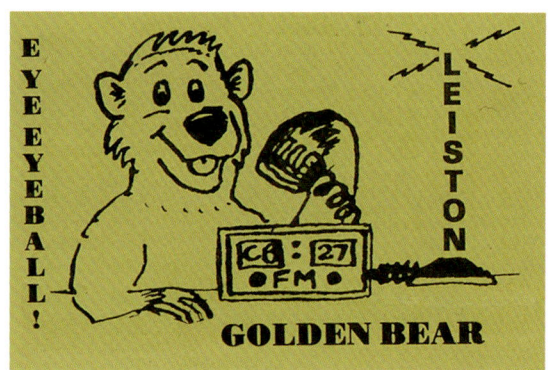

CONGRATULATIONS!
YOU HAVE JUST EYEBALLED

DOUBLE TROUBLE

Wickham Market Breaker

10-10 TILL WE BREAK AGAIN

EYEBALL! EYEBALL!
You Have Just Eyeballed

Racer & Knitting Needles

Chicken Town Breakers
10-10 Till We Race Again

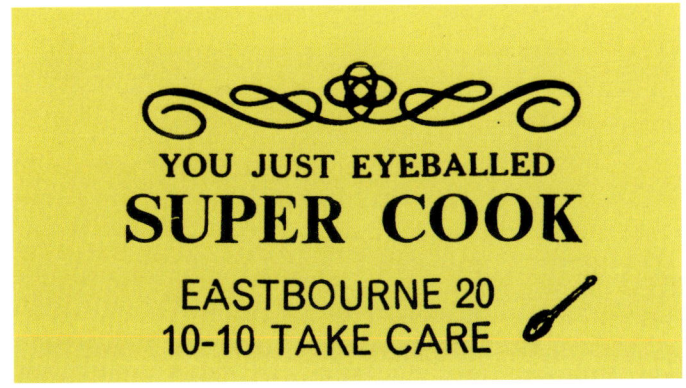

YOU JUST EYEBALLED
SUPER COOK
EASTBOURNE 20
10-10 TAKE CARE

EYEBALL! EYEBALL!
YOU HAVE JUST EYEBALLED

★ ★ ★ ★ ★ ★ ★ ★ ★

LITTLE PLUM

★ ★ ★ ★ ★ ★ ★ ★ ★

10 - 10 Till we EYEBALL again

You're being
BEWITCHED
By Blythe
Spirit
WHITTON 20

You Have Just Eye-Balled

MIDNIGHT
DRIFTER

Bexhill Twenty
(Foggy Town)

73s   Norman   88s

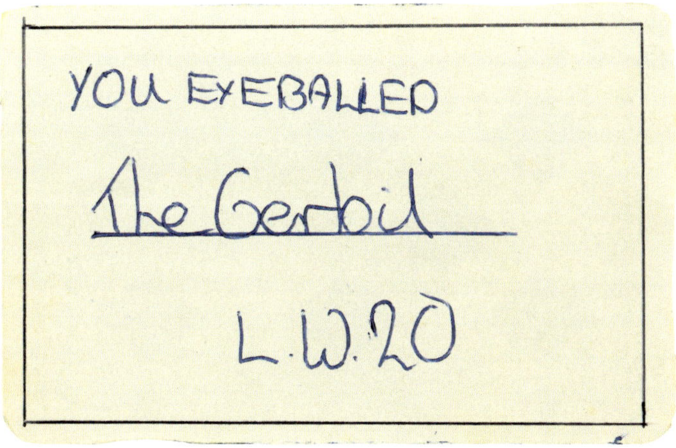

YOU EYEBALLED

The Gerbil

L.W. 20

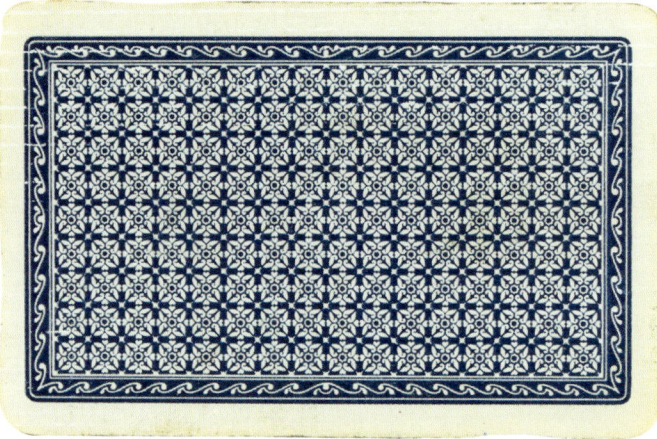

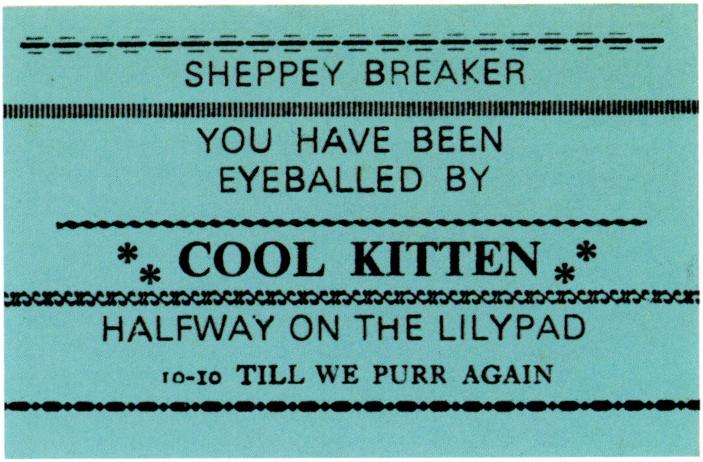

SHEPPEY BREAKER

YOU HAVE BEEN
EYEBALLED BY

** **COOL KITTEN** **

HALFWAY ON THE LILYPAD

10-10 TILL WE PURR AGAIN

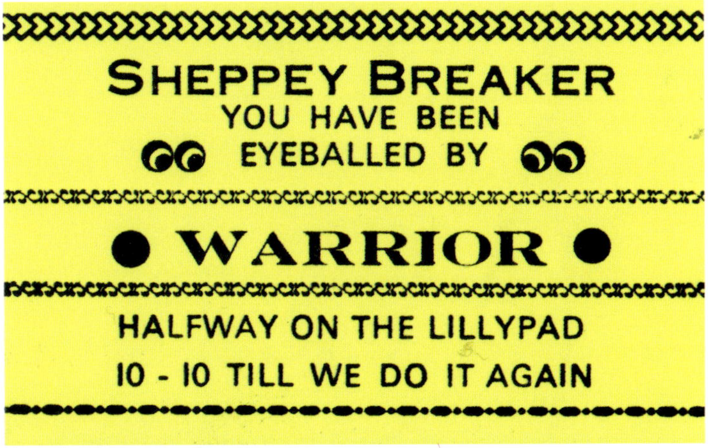

SHEPPEY BREAKER
YOU HAVE BEEN
👀 EYEBALLED BY 👀

● **WARRIOR** ●

HALFWAY ON THE LILLYPAD

10 - 10 TILL WE DO IT AGAIN

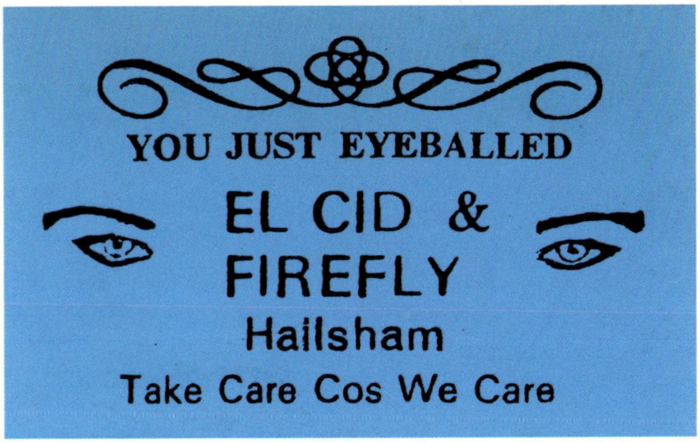

YOU JUST EYEBALLED

**EL CID &
FIREFLY**

Hailsham

Take Care Cos We Care

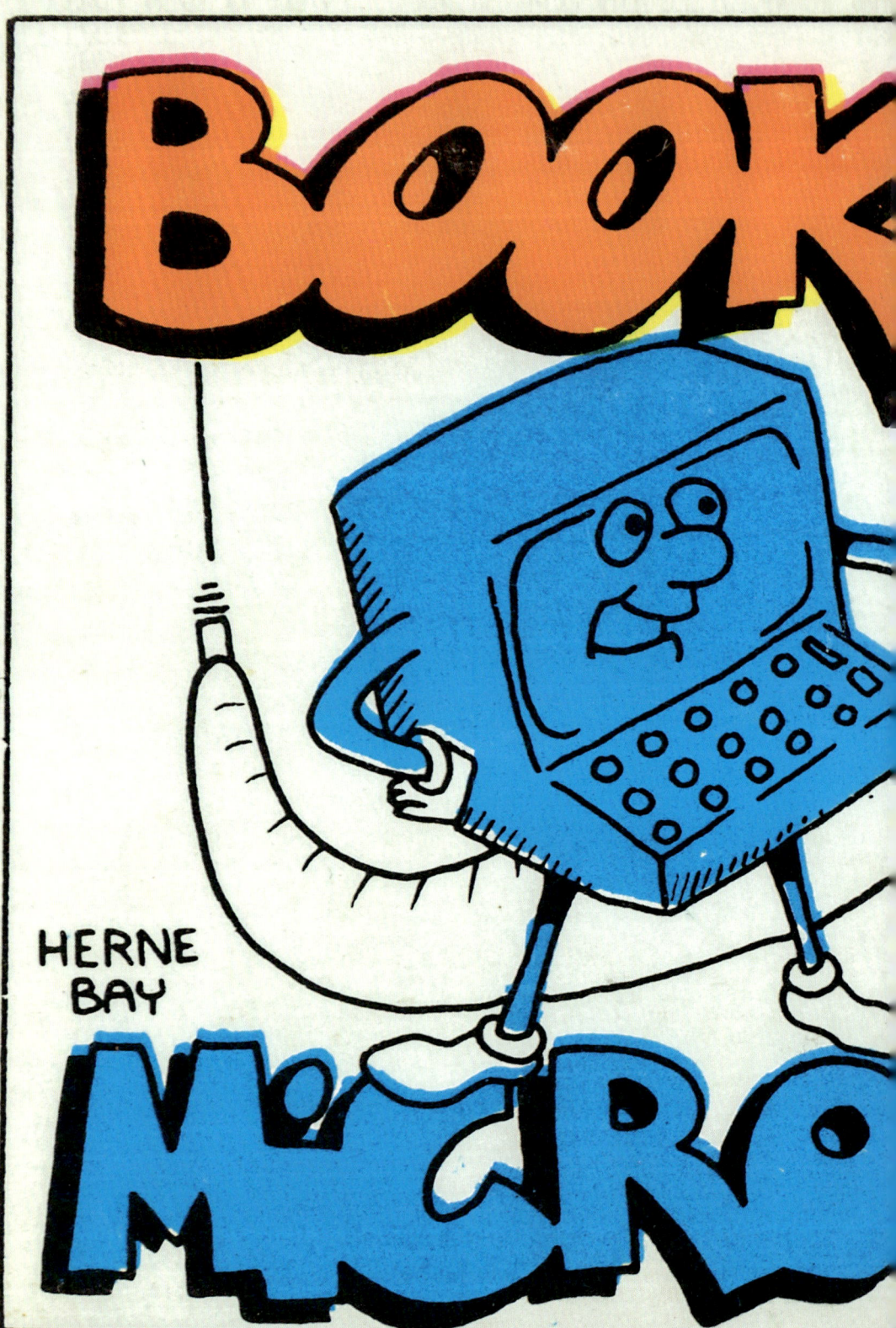

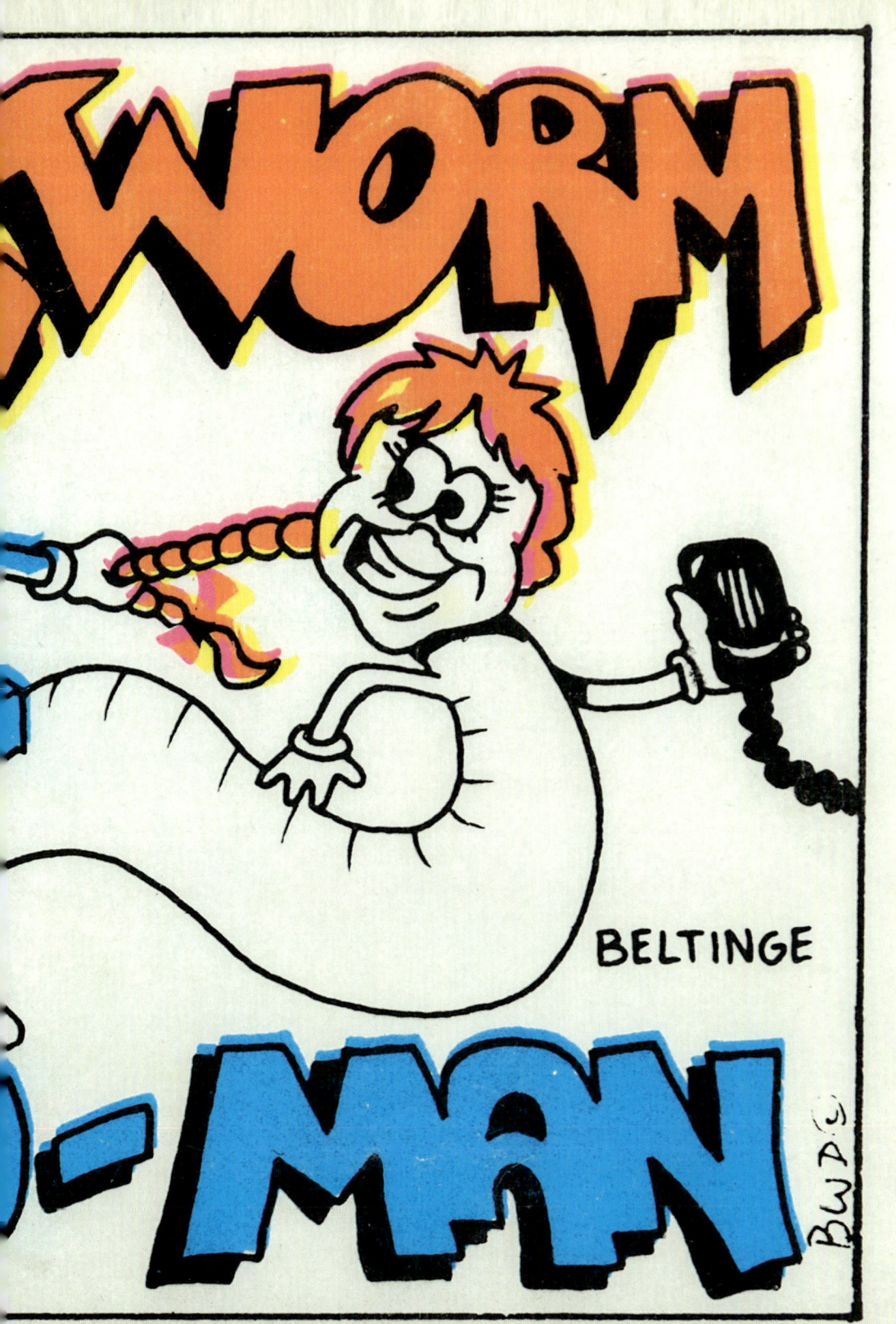

CONGRATULATIONS!
YOU HAVE JUST EYEBALLED

FRUITCAKE

Sizewell Breaker

10-10 TILL WE BREAK AGAIN

Jenny Dye, aka 'Fruitcake'.

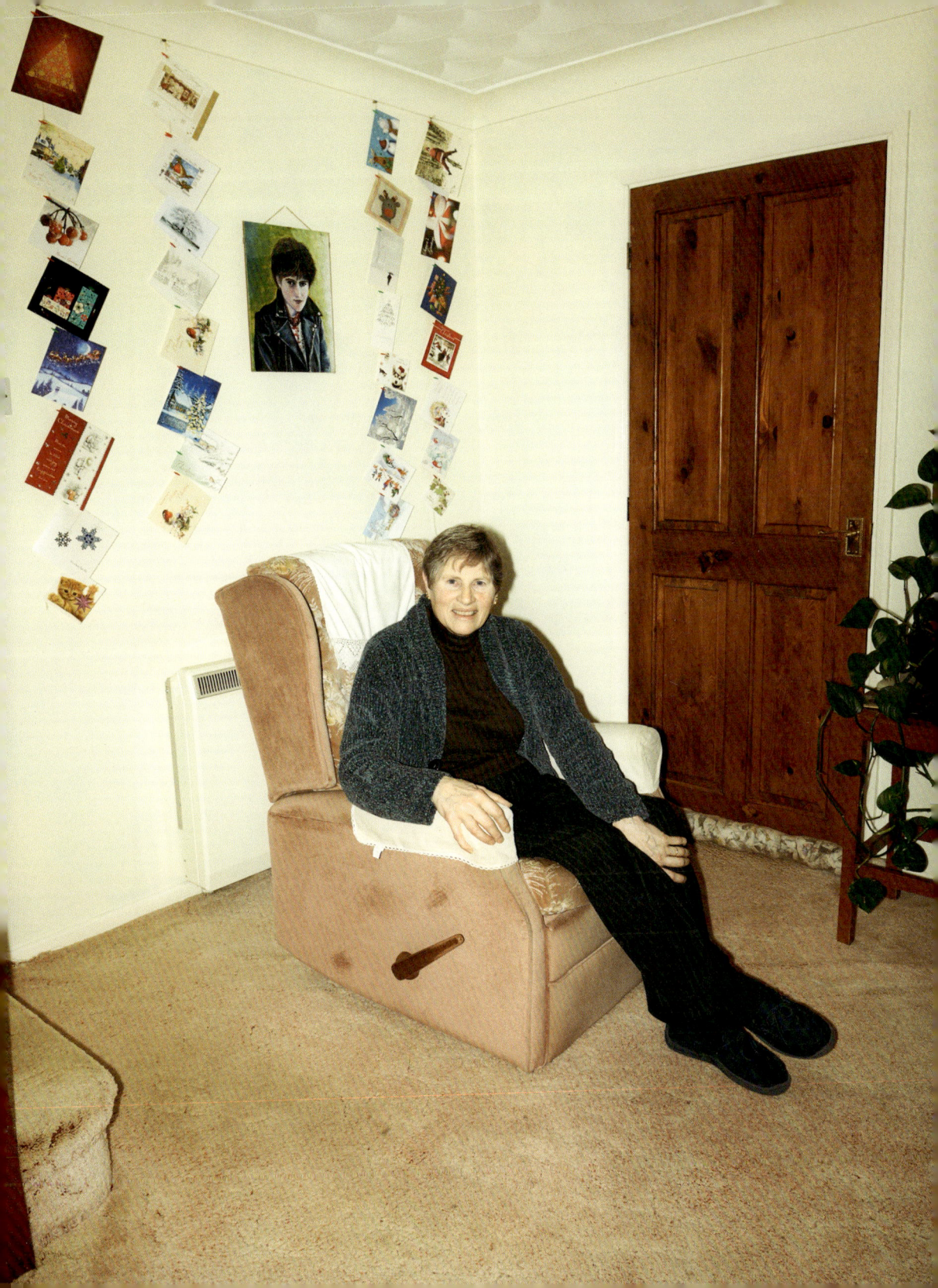

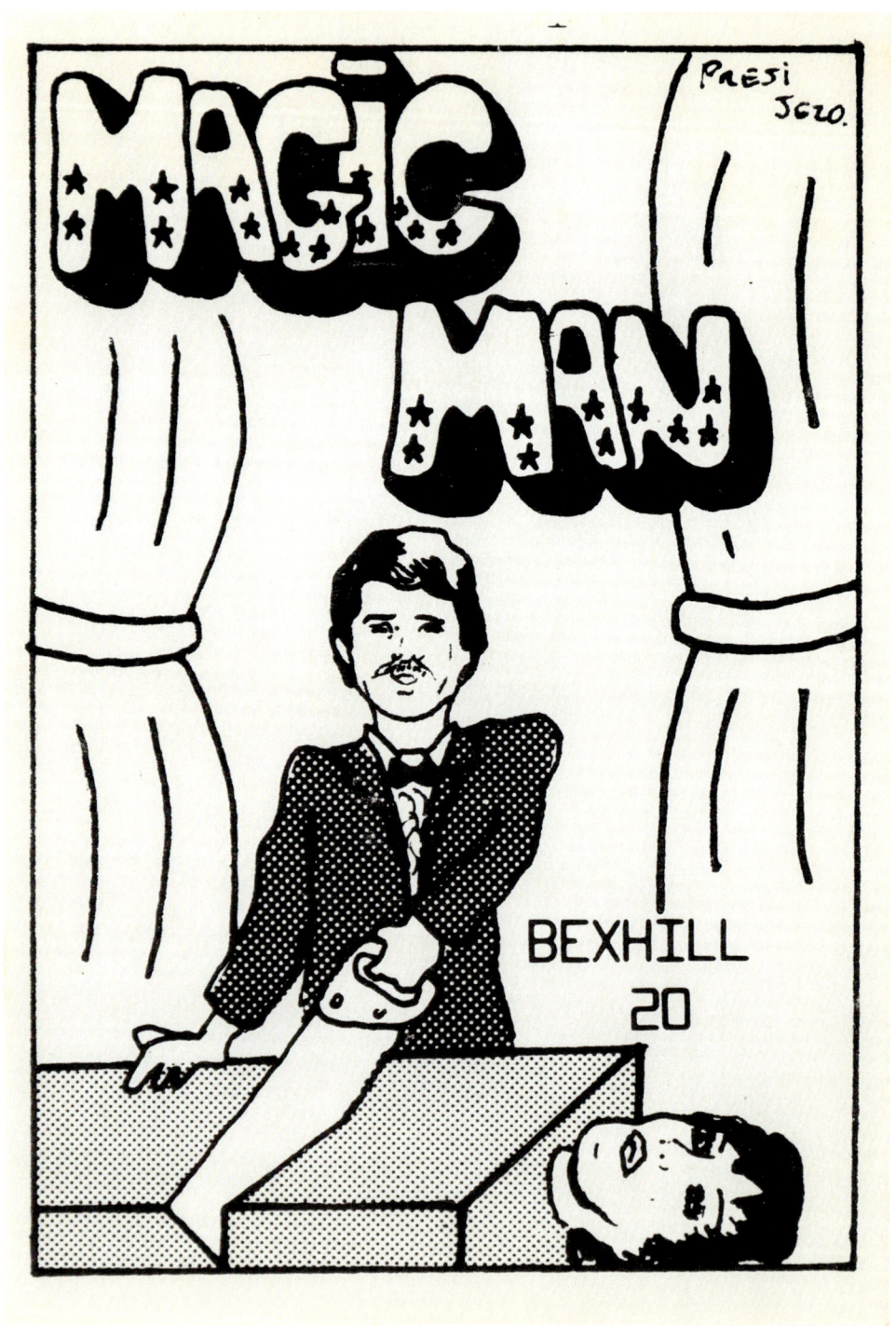

You Have Just Eye-Balled

RIZLA

88s   Newtown   73s

Twenty

10-10 Till We Do It Again

EYEBALL! EYEBALL!

You Have Just Eyeballed

FERRETER

Mistley Breaker

10-10 Till We Break Again

EYEBALL !   EYEBALL !

You Have Just Eyeballed

IRONMAN

Saxmundham   Twenty

10 - 10 Till we Meet again

YOU HAVE JUST EYEBALLED

MAID MARIAN

and

FRIAR TUCK

BECKENHAM 20

TAKE CARE COS WE CARE

It's better to have Eyeballed Colin

The KNEECAPPER

than to have upset him

SYDENHAM 20

TAKE CARE COS WE CARE

DANGER Government Health WARNING
KNEECAPPER CAN
SERIOUSLY DAMAGE YOUR HEALTH

EYEBALL   EYEBALL

You Have Just Eyeballed

★★★★★★★★★★★★★★★★★

Little Mog

★★★★★★★★★★★★★★★★★

CASTLE TOWN 25 CLUB
10 - 10 Till we MOGULATE again

CONGRATULATIONS!

YOU HAVE JUST EYEBALLED

WELLINGTON

Yoxford Breaker

10-10 TILL WE BREAK AGAIN

YOU'VE EYEBALLED

ROLLERBALL

BEXHILL 20

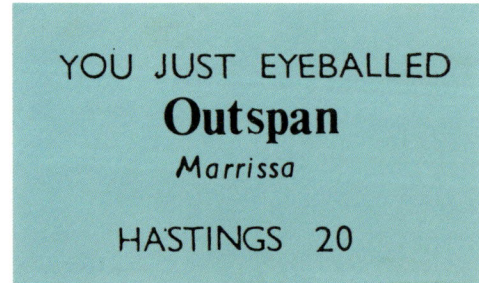

YOU JUST EYEBALLED
Outspan
Marrissa

HASTINGS 20

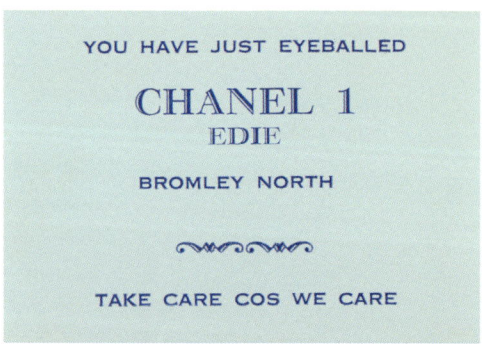

YOU HAVE JUST EYEBALLED

CHANEL 1
EDIE

BROMLEY NORTH

TAKE CARE COS WE CARE

You have just been
Eyeballed by

SKINNY RIBS
SUNDRIDGE PARK

10–10 till we meet again

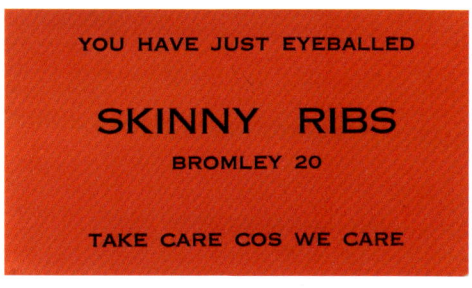

YOU HAVE JUST EYEBALLED

SKINNY RIBS
BROMLEY 20

TAKE CARE COS WE CARE

EYEBALL! EYEBALL!
You Have Just Eyeballed

RED AMIGO

10-10 Till We Break Again

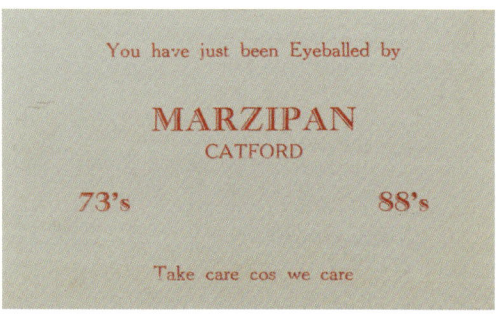

You Have Eyeballed

Silver Dog
Peter
Hastings 20

You have just been
Eyeballed by

BABY DOLL
SUNDRIDGE PARK

10–10 till we meet again

You have just been Eyeballed by

MARZIPAN
CATFORD

73's   88's

Take care cos we care

46

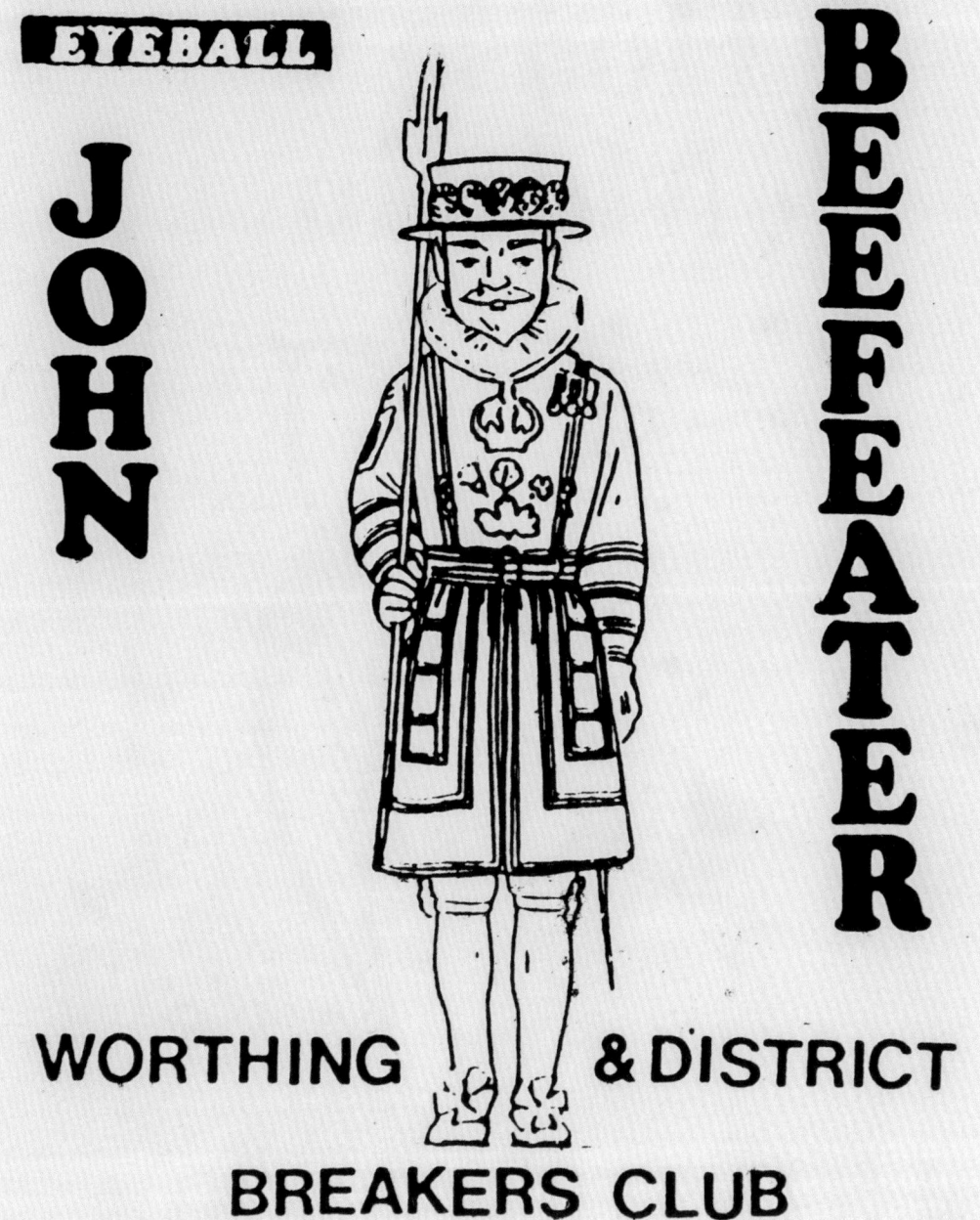

EYEBALL

JOHN

BEEFEATER

WORTHING &DISTRICT

BREAKERS CLUB

House
Martin

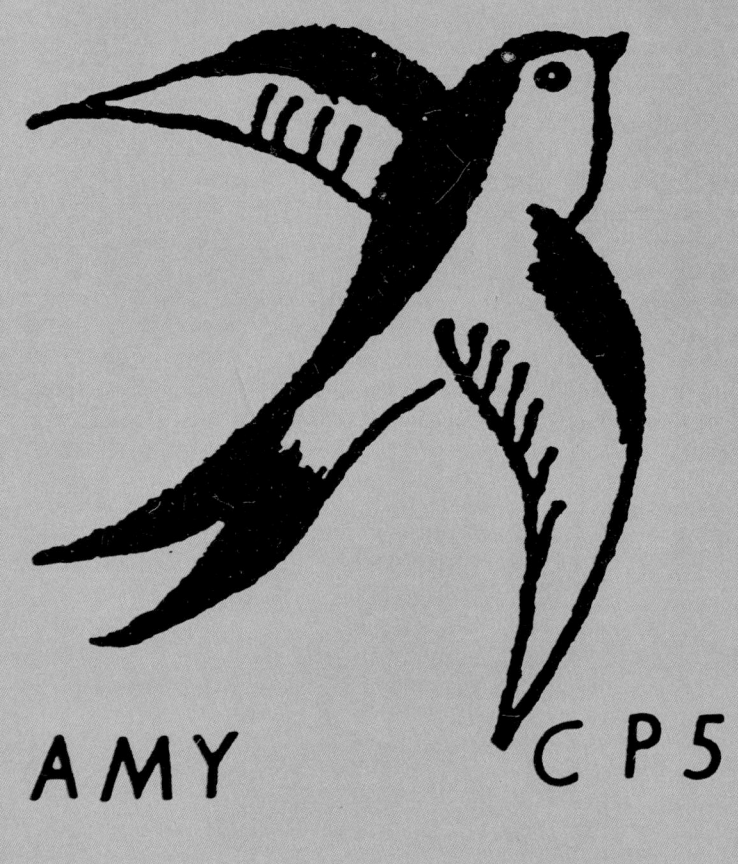

AMY   C P 5

HASTINGS   20

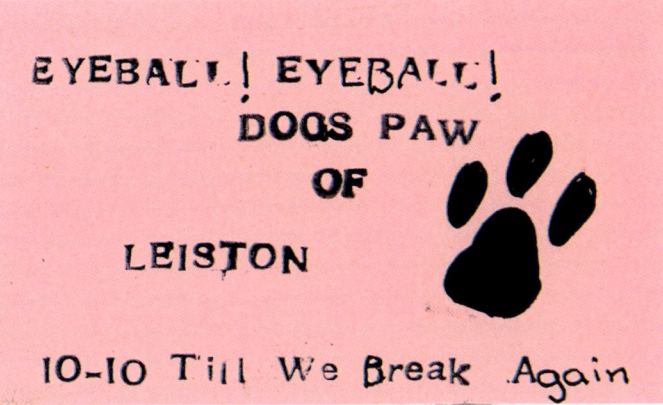

EYEBALL! EYEBALL!
DOGS PAW
OF
LEISTON

10-10 Till We Break Again

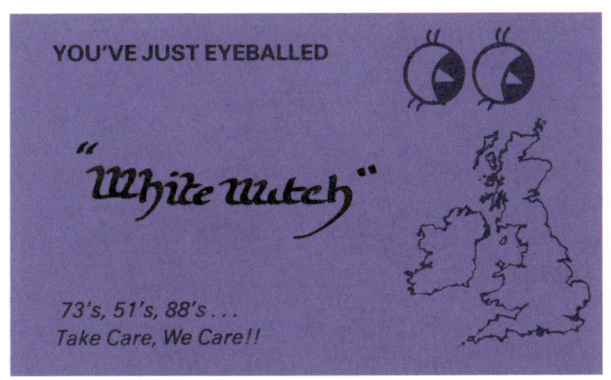

YOU'VE JUST EYEBALLED

"White Watch"

73's, 51's, 88's . . .
Take Care, We Care!!

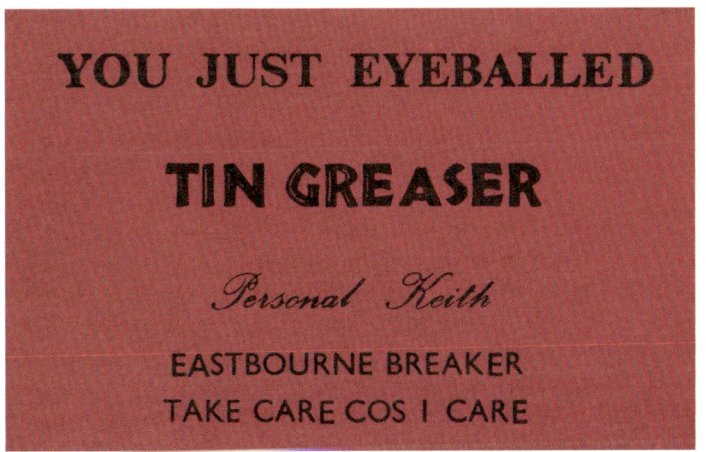

YOU JUST EYEBALLED

TIN GREASER

Personal Keith

EASTBOURNE BREAKER
TAKE CARE COS I CARE

GERMAN

SHEPHERD

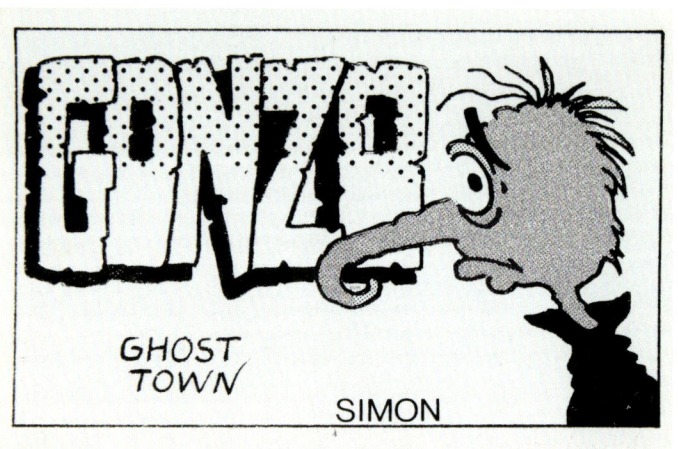

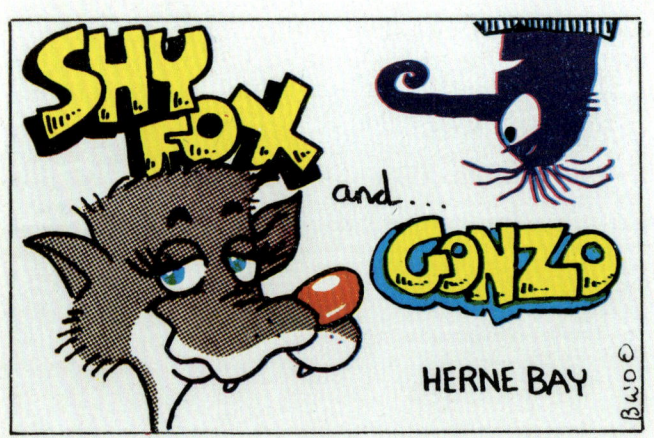

You Have
Eyeballed

JOE SOAP

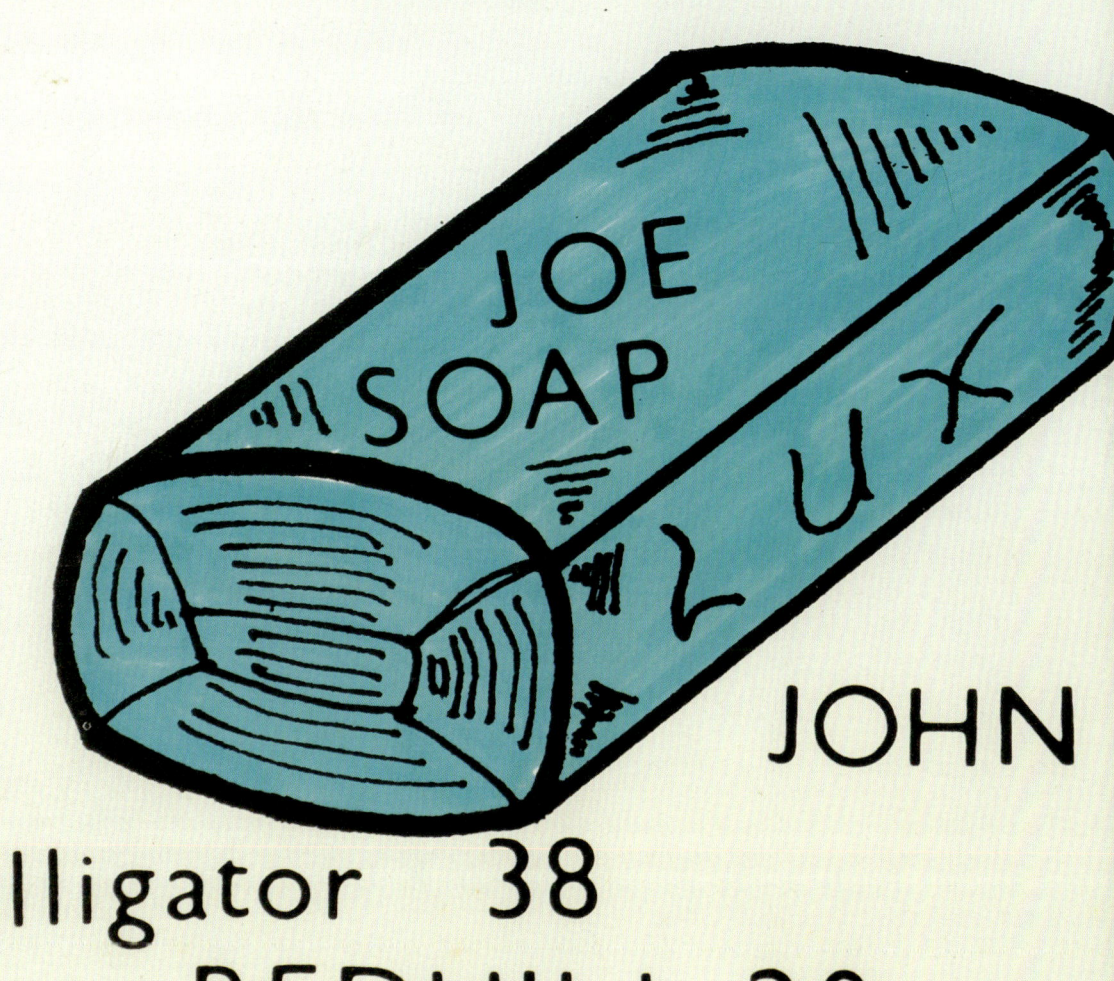

JOHN

Alligator 38

REDHILL 20

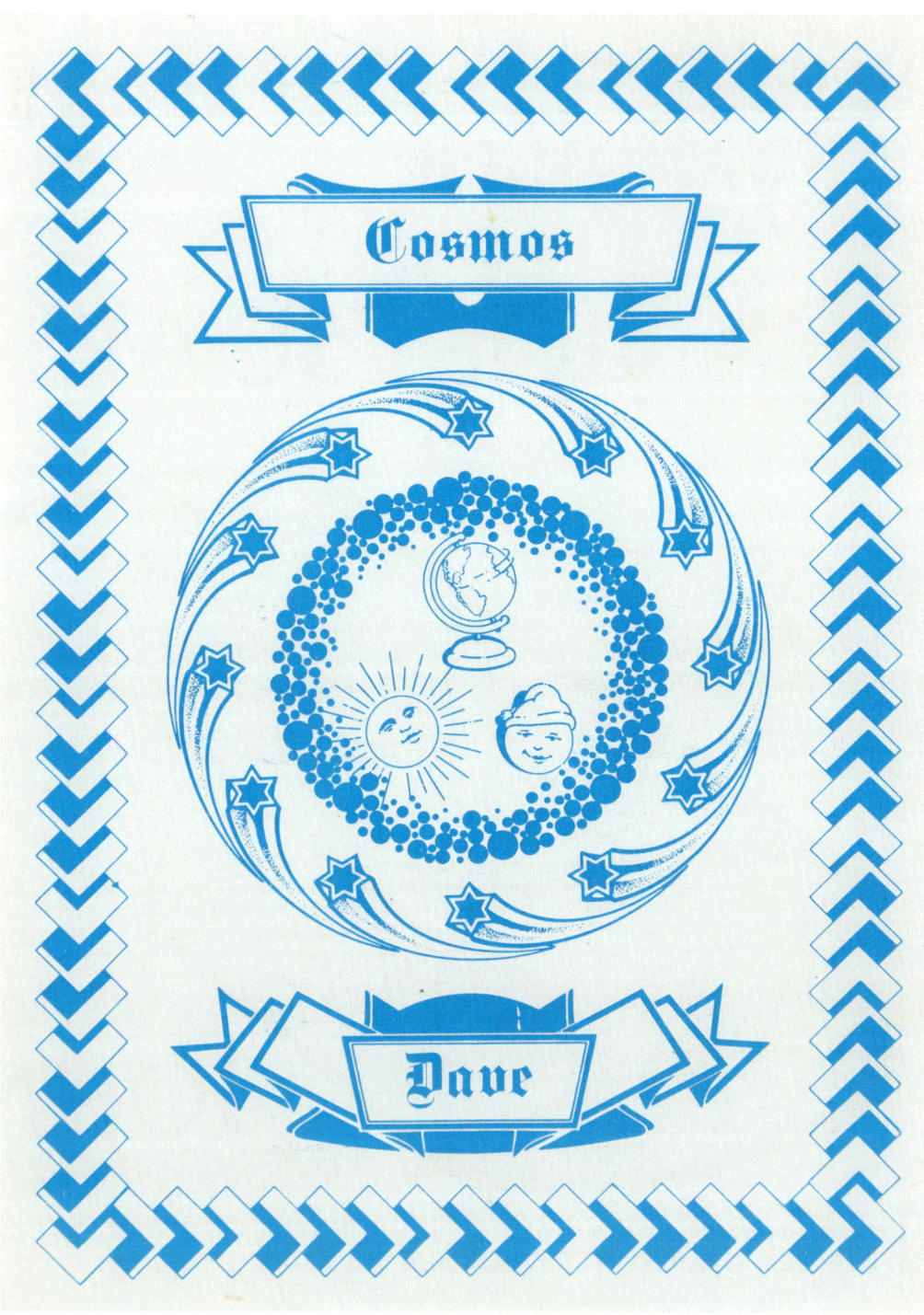

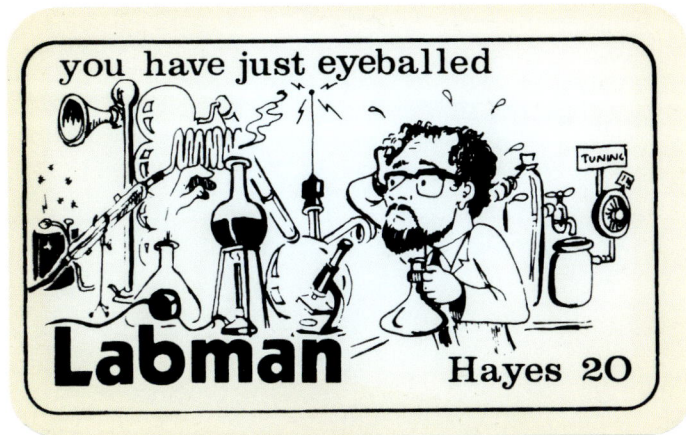

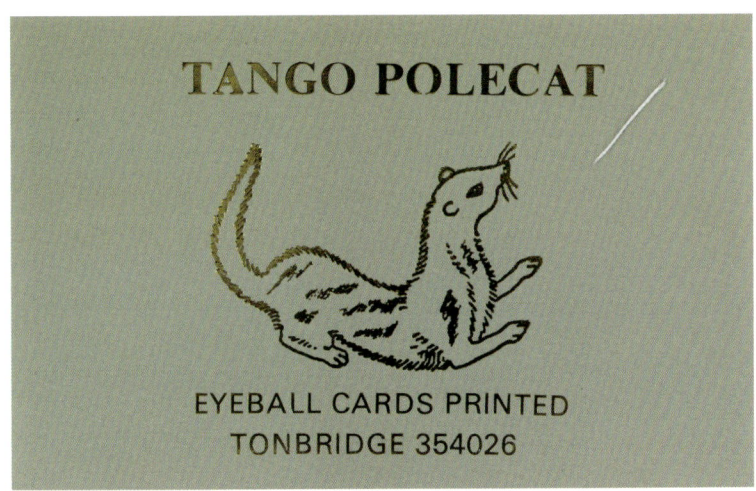

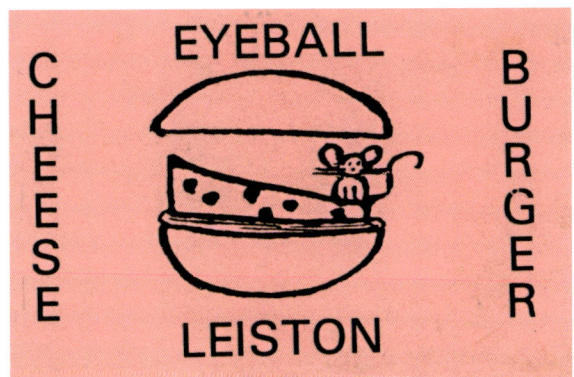

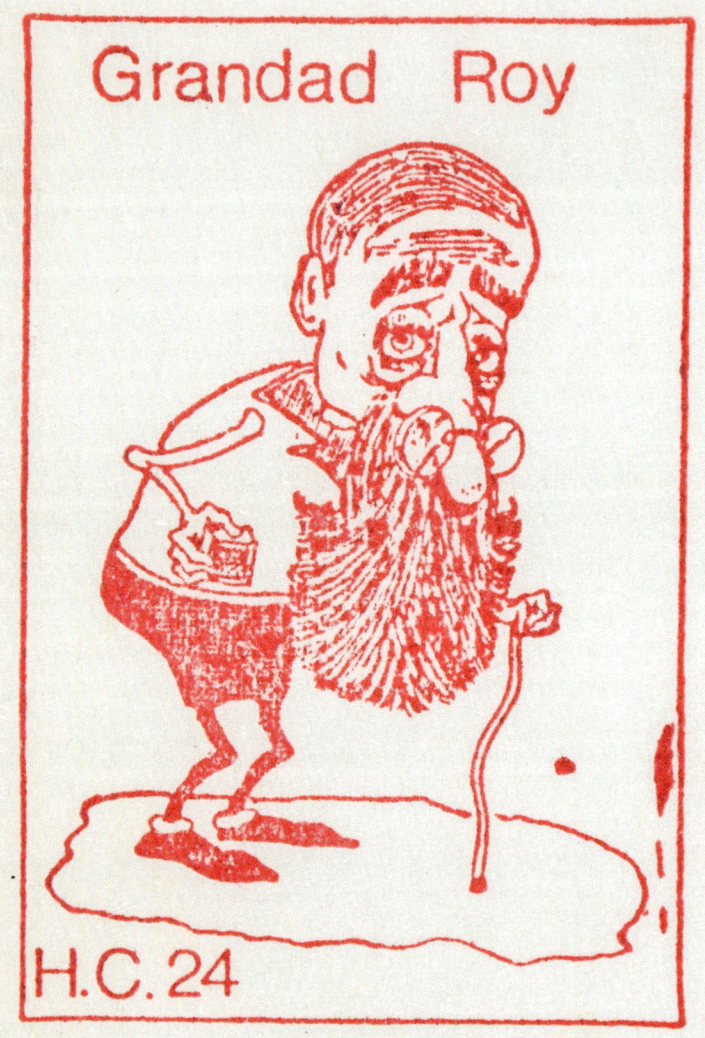

Grandad Roy

H.C. 24

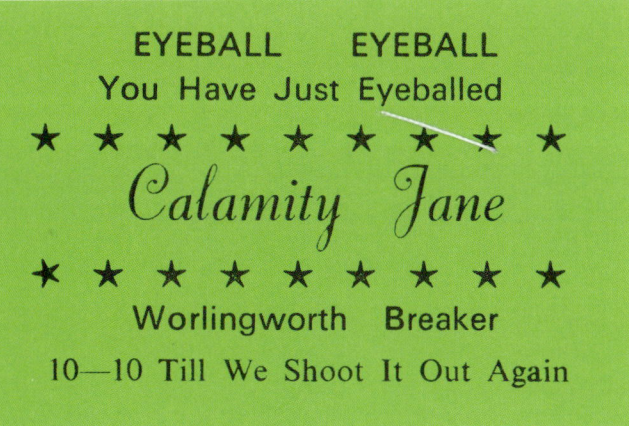

EYEBALL   EYEBALL
You Have Just Eyeballed
★ ★ ★ ★ ★ ★ ★ ★ ★ ★
Calamity Jane
★ ★ ★ ★ ★ ★ ★ ★ ★ ★
Worlingworth   Breaker
10—10 Till We Shoot It Out Again

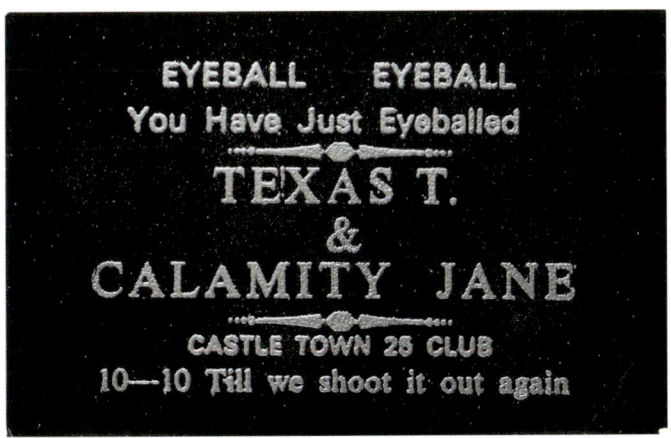

EYEBALL !   EYEBALL !
YOU HAVE JUST EYEBALLED

TEXAS T
CASTLE TOWN 25 CLUB

10—10 Till we shoot it out again

EYEBALL   EYEBALL
You Have Just Eyeballed
TEXAS T.
&
CALAMITY JANE
CASTLE TOWN 25 CLUB
10—10 Till we shoot it out again

FLINT MAN

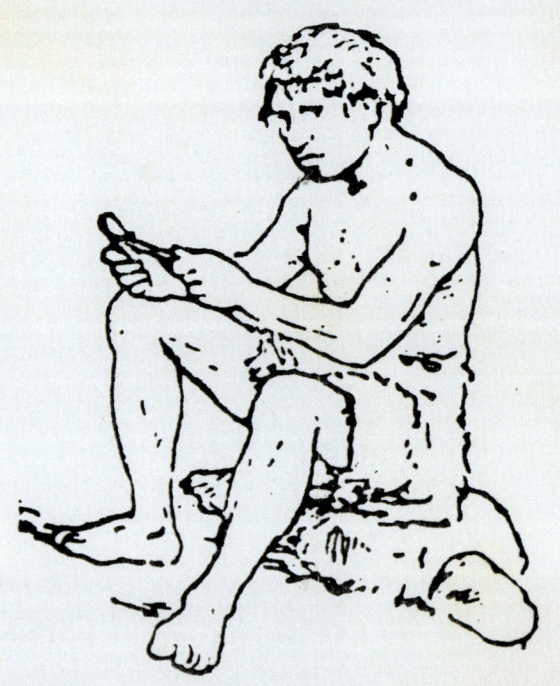

Hastings

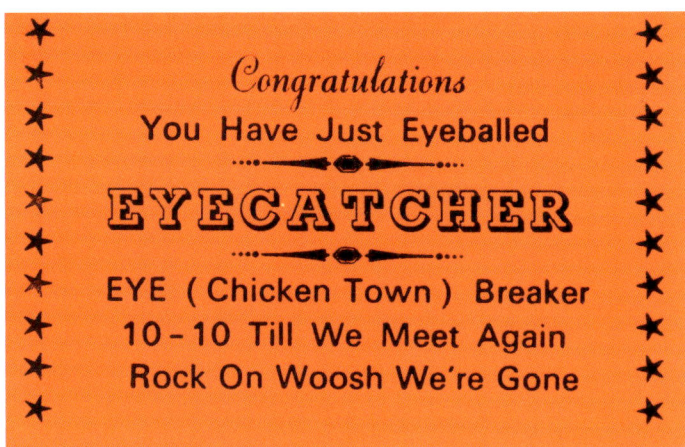

Congratulations
You Have Just Eyeballed

EYECATCHER

EYE (Chicken Town) Breaker
10 - 10 Till We Meet Again
Rock On Woosh We're Gone

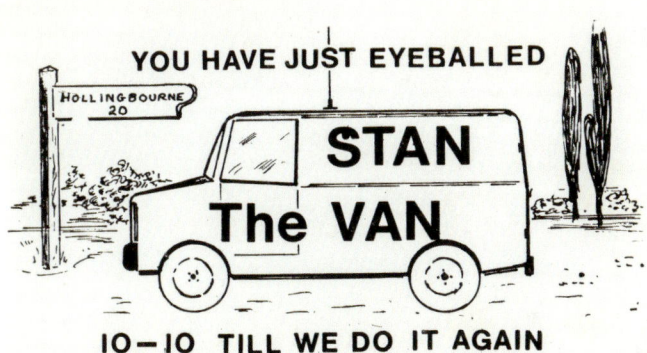

YOU HAVE JUST EYEBALLED

HOLLINGBOURNE 20

STAN
The VAN

10 – 10 TILL WE DO IT AGAIN

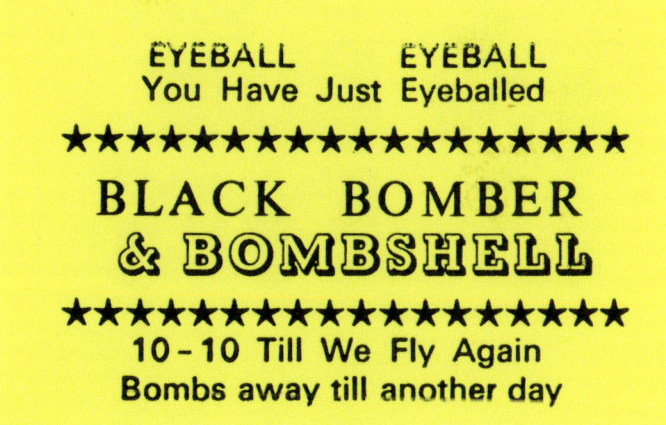

EYEBALL   EYEBALL
You Have Just Eyeballed

★★★★★★★★★★★★★★★★★★

BLACK BOMBER
& BOMBSHELL

★★★★★★★★★★★★★★★★★★

10 - 10 Till We Fly Again
Bombs away till another day

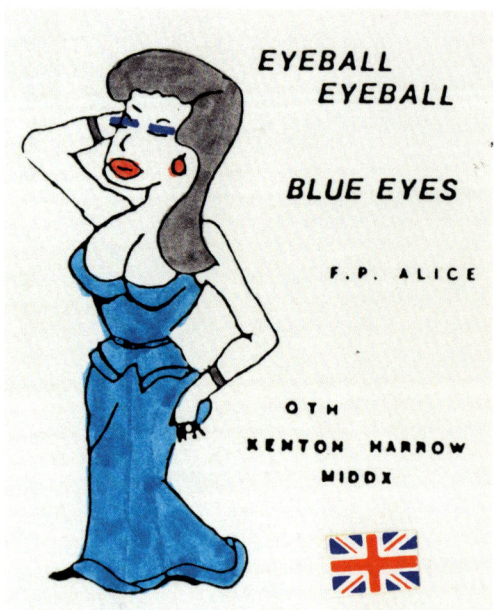

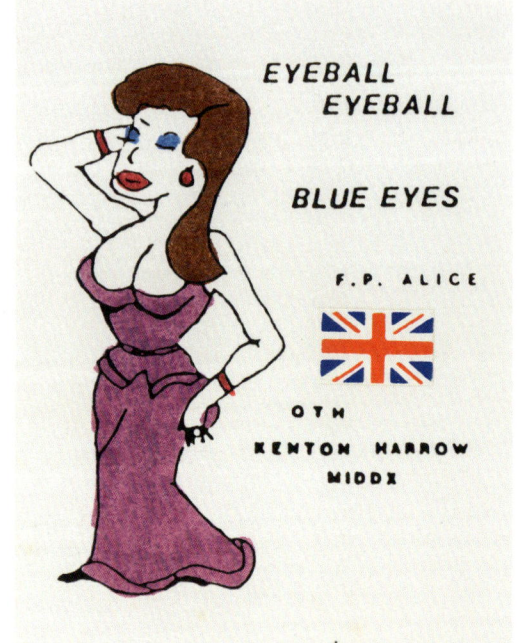

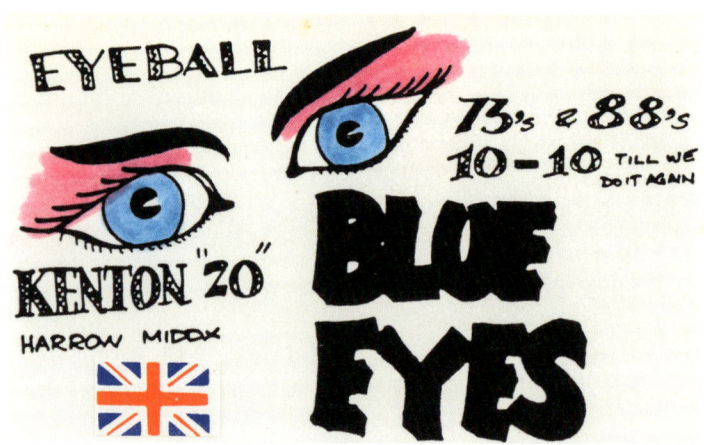

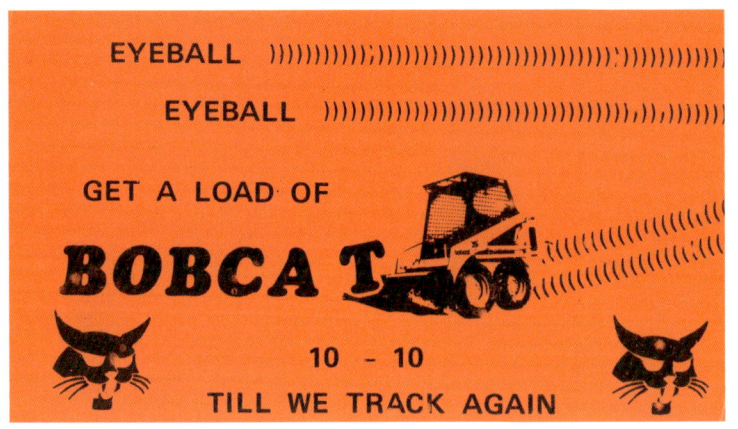

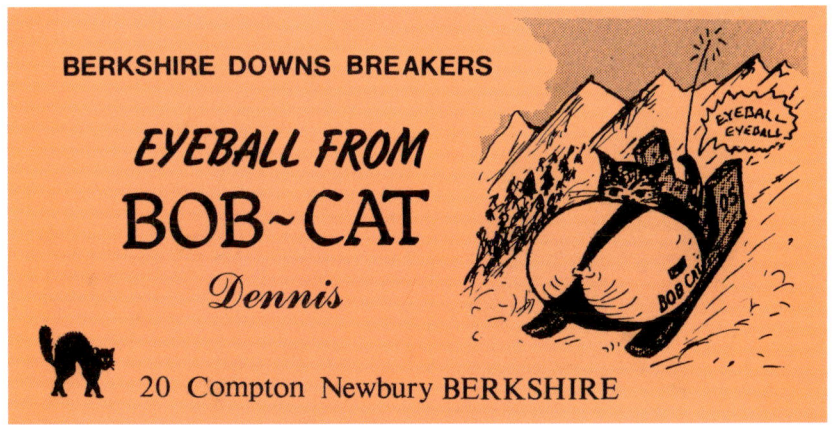

EYEBALL! EYEBALL!
You Have Just Eyeballed

WOOLY BULLY & DINKY DORA

10-10 Till We Break Again

EYEBALL! EYEBALL!
You Have Just Eyeballed

SPRAY CAN

10-10 Till We Break Again

EYEBALL! EYEBALL!
You Have Just Eyeballed

LITTLE F

Bedingfield Breaker

10-10 Till We Do It Again

Woking 20
You've Just Eyeballed

ACKER BILK
(Roy)

10-10 Till We Meet Again

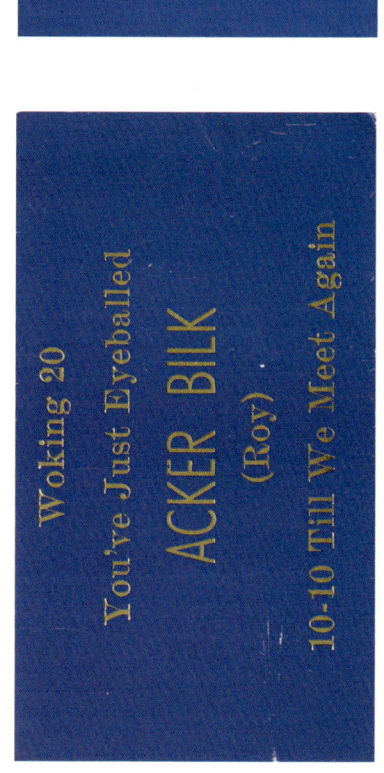

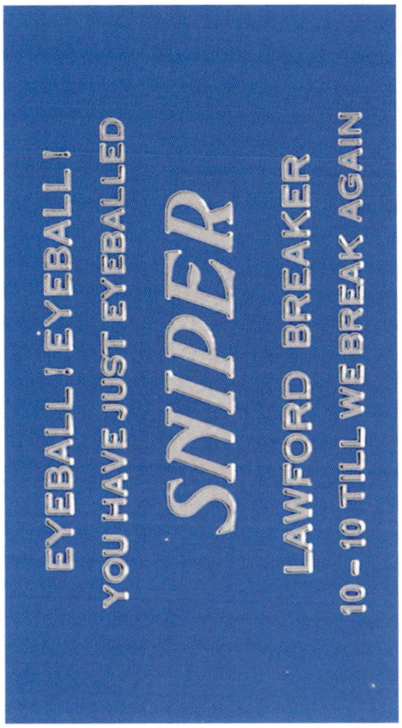

EYEBALL! EYEBALL!

YOU HAVE JUST EYEBALLED

SNIPER

LAWFORD BREAKER

10 – 10 TILL WE BREAK AGAIN

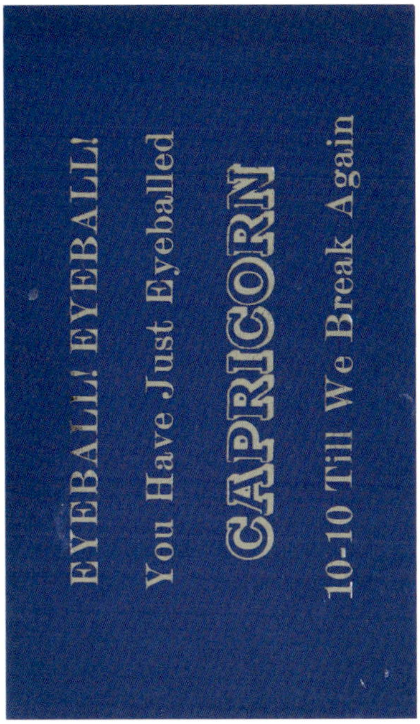

EYEBALL! EYEBALL!

You Have Just Eyeballed

CAPRICORN

10-10 Till We Break Again

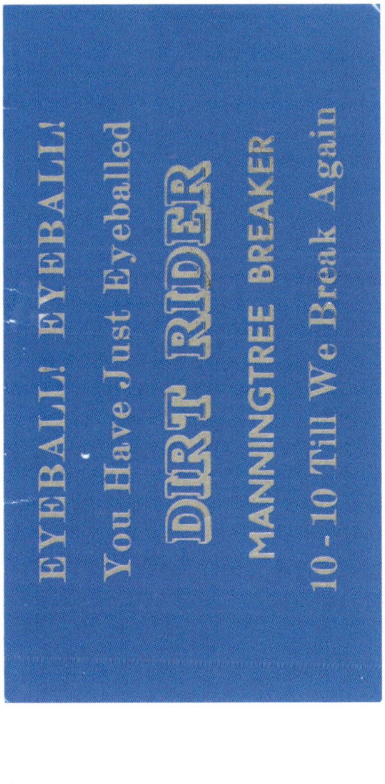

EYEBALL! EYEBALL!

You Have Just Eyeballed

DIRT RIDER

MANNINGTREE BREAKER

10 - 10 Till We Break Again

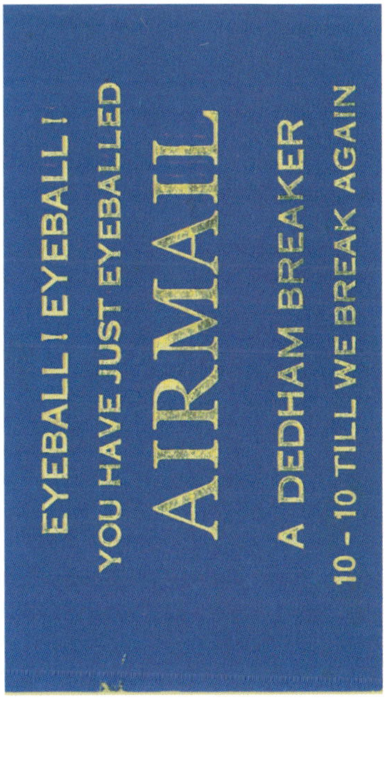

EYEBALL! EYEBALL!

YOU HAVE JUST EYEBALLED

AIRMAIL

A DEDHAM BREAKER

10 - 10 TILL WE BREAK AGAIN

EYEBALL! EYEBALL!
You've Just Eyeballed

**PICKLES &
FLYING HERDSMAN**

10-10 Till We Meet Again

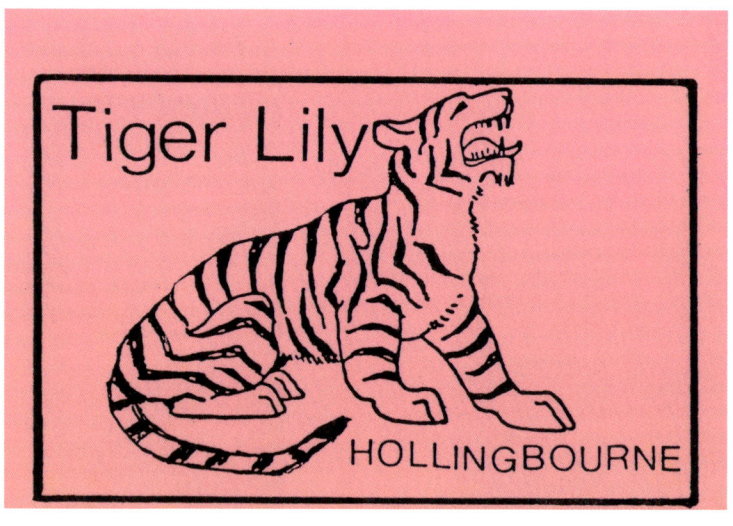

Tiger Lily

HOLLINGBOURNE

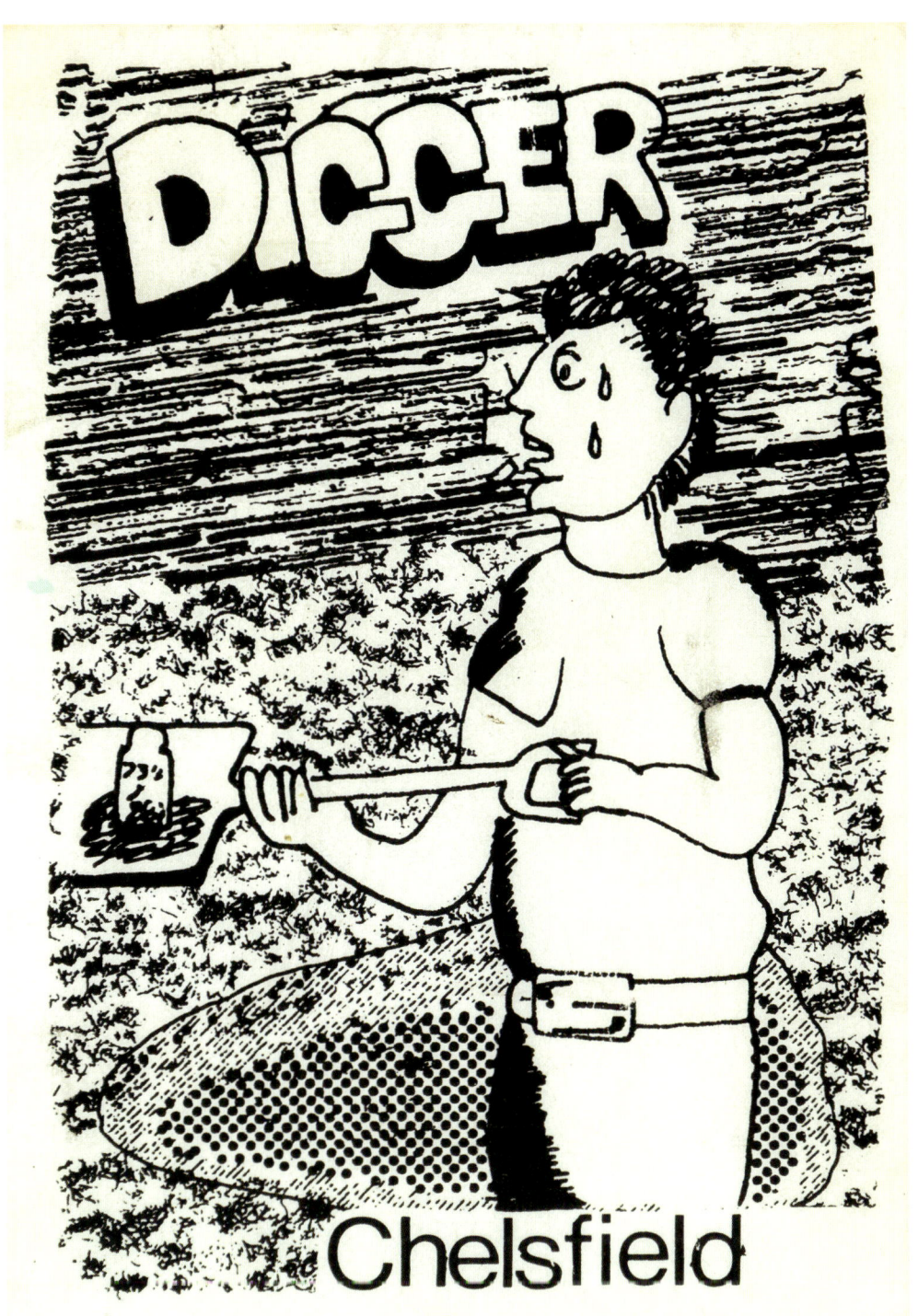

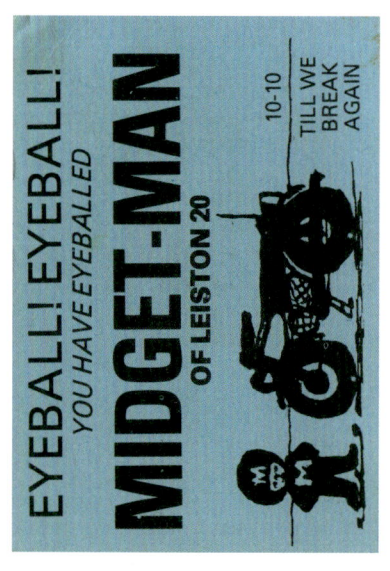

Mark Cracknell, aka 'Midget Man'.

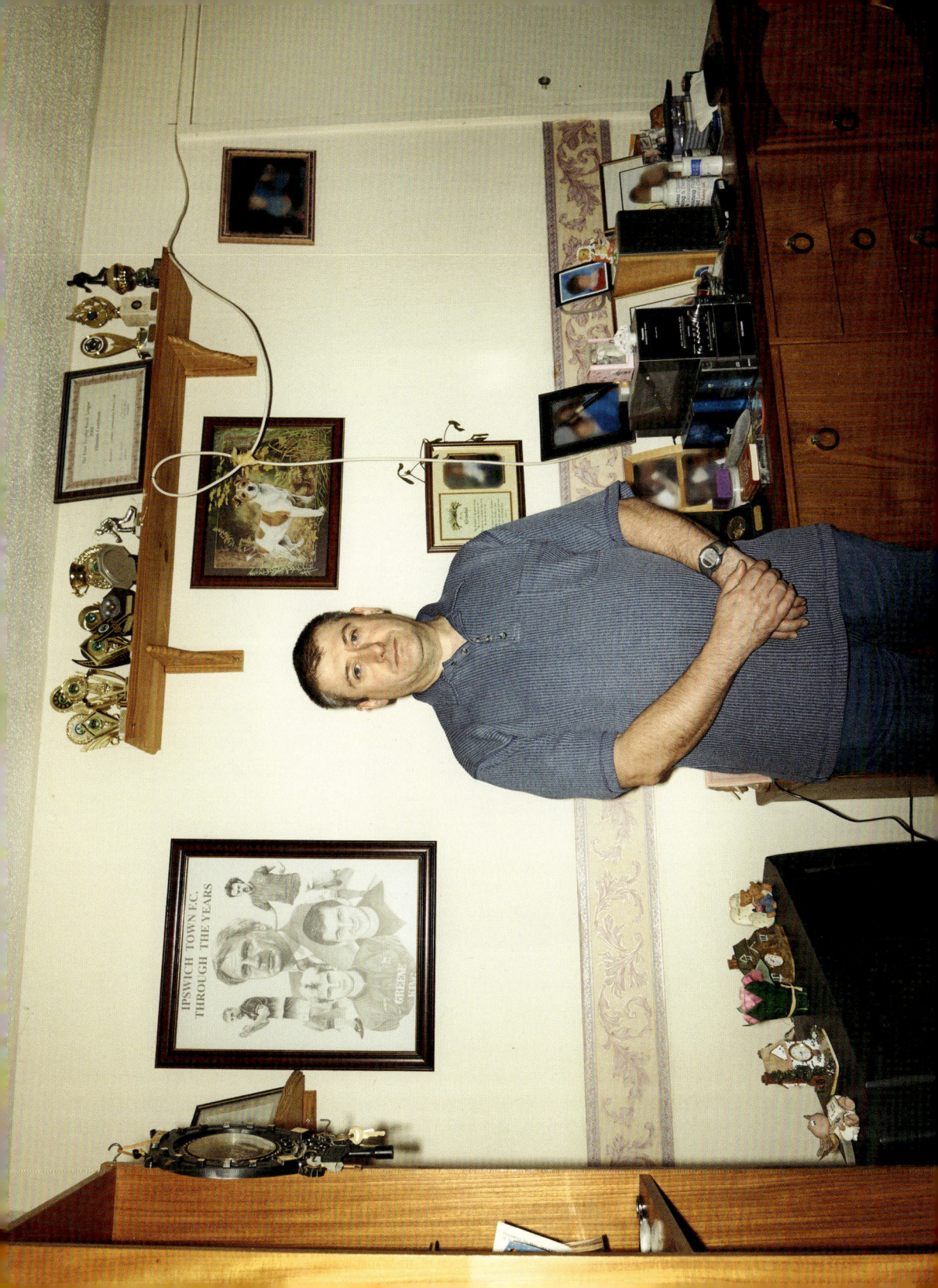

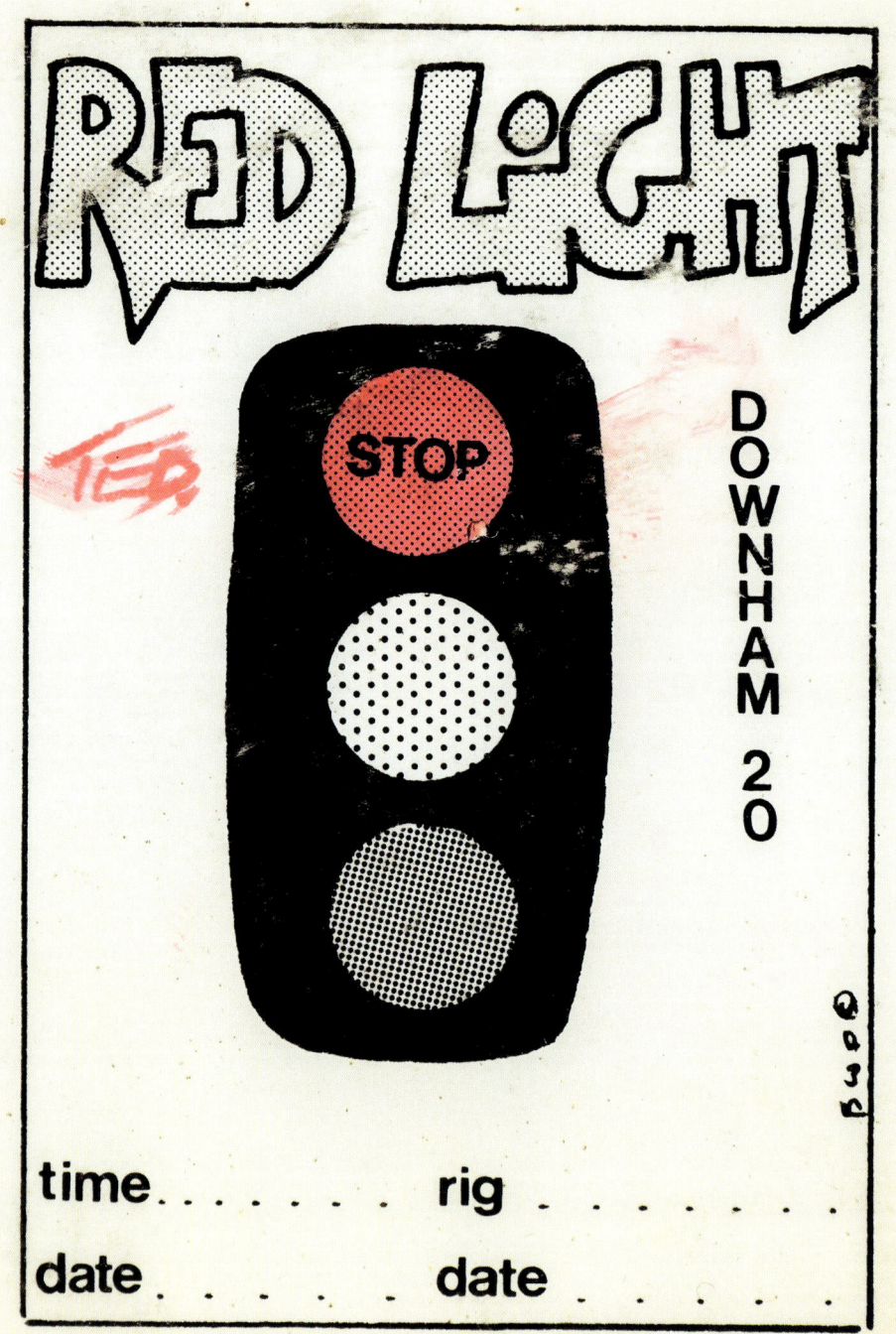

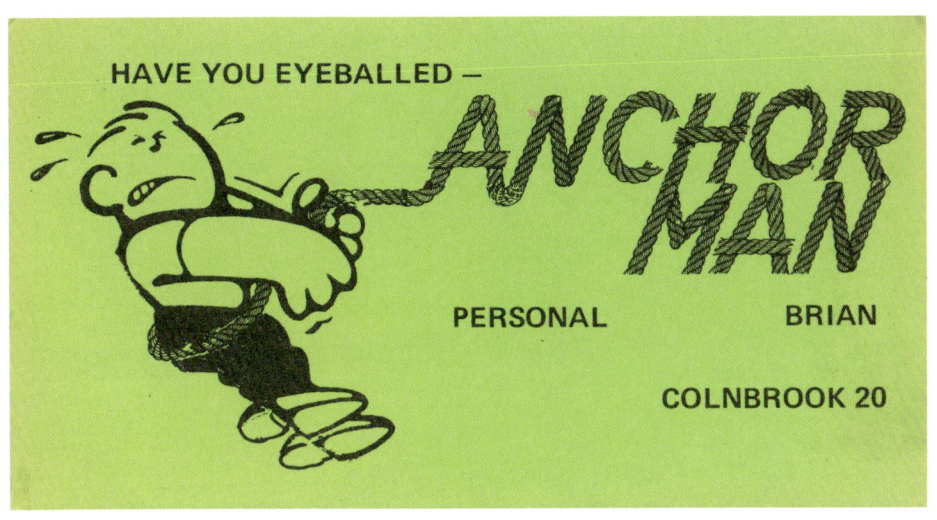

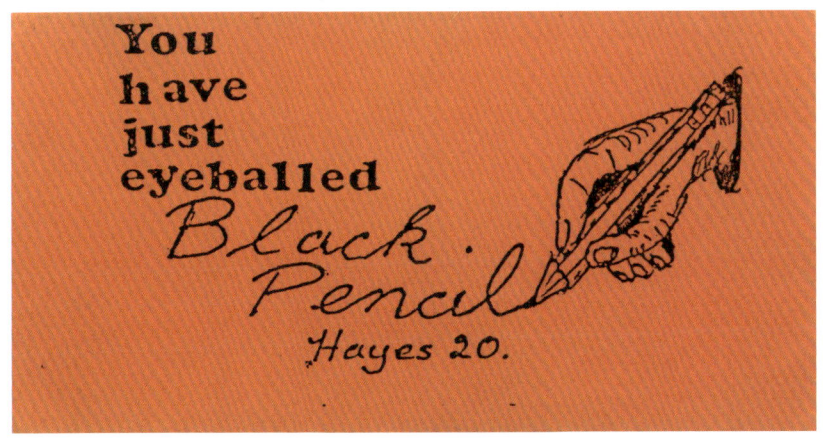

EYEBALL!

DUKE

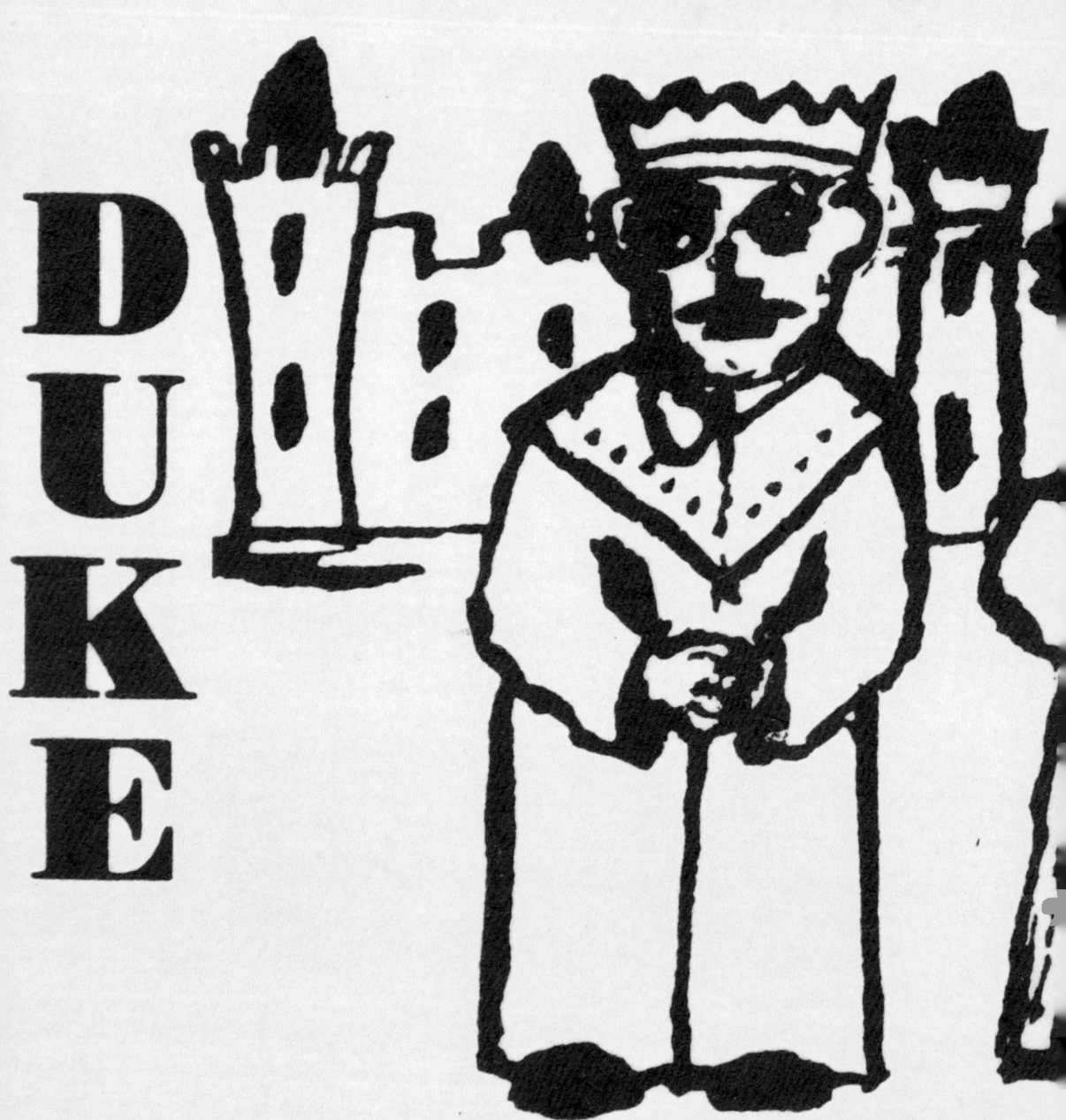

KELSA

EYEBALL!

DUCHESS

LE 20

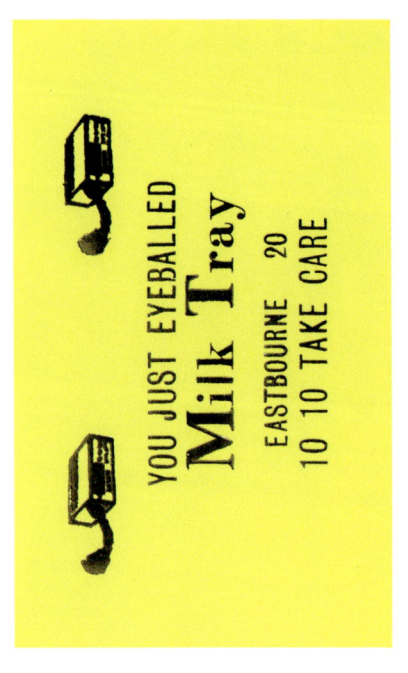

YOU JUST EYEBALLED
Milk Tray
EASTBOURNE 20
10 10 TAKE CARE

EYEBALL   EYEBALL
You Have Just Eyeballed
★ ★ ★
★ ★ ★
Curly
★ ★ ★
★ ★ ★
★ ★ ★
Framlingham 25 Breakers Club
10 – 10 Till We Meet Again

EYEBALL   EYEBALL
You Have Just Eyeballed

Mighty Mo

★
★ ★
10 – 10   T. T. F. N.
WE UP WE DOWN
WE'RE GONE

Maxi Man
BRIAN

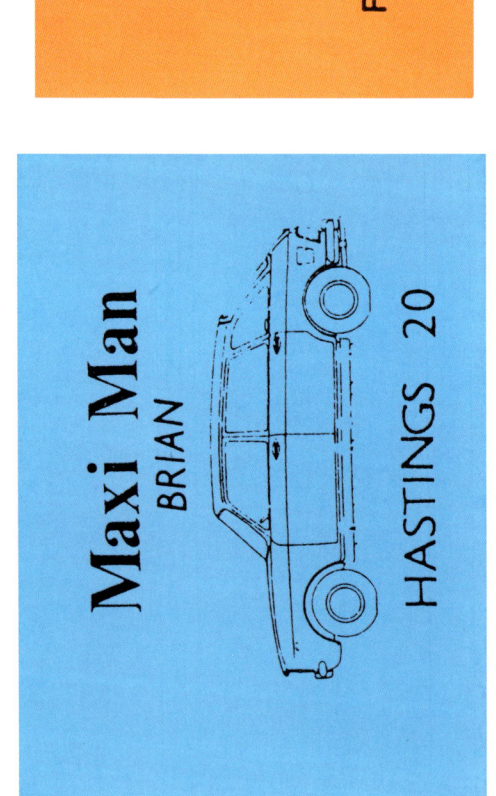

HASTINGS 20

You have just been Eyeballed by

MARATHON LADY
CATFORD

10-10 Till we do it again

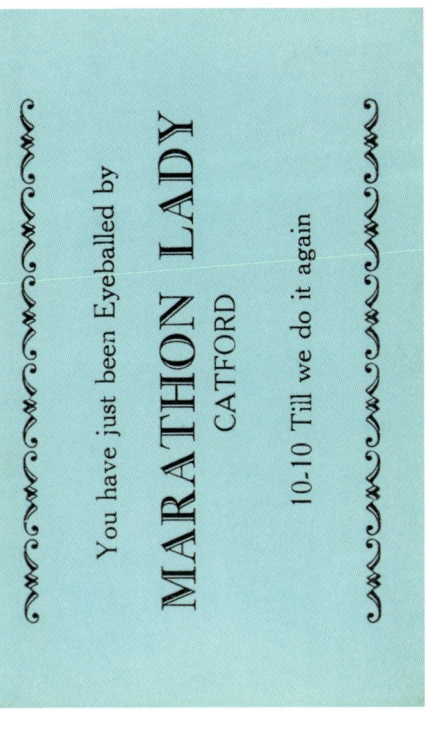

YOU HAVE JUST EYEBALLED

S I L V E R...S T R E A K

WIMBLEDON
2 0

IO IO TILL WE MEET AGAIN

EYEBALL! EYEBALL!
YOU HAVE JUST EYEBALLED

COMBINE HARVESTER
DENNINGTON BREAKER

10-10 Till The Corn Comes Again

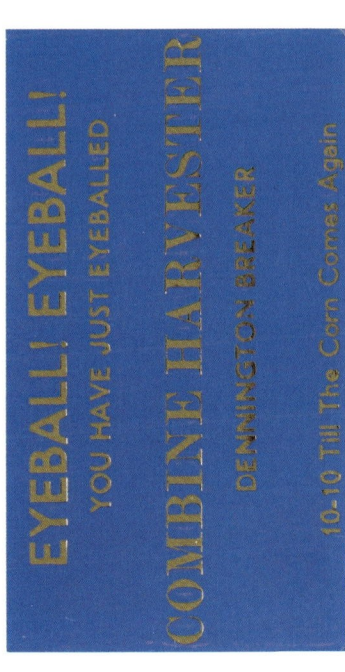

EYEBALL! EYEBALL!
YOU HAVE JUST EYEBALLED

COWBOY
CASTLE TOWN 25 CLUB

10—10 Till we shoot it out again

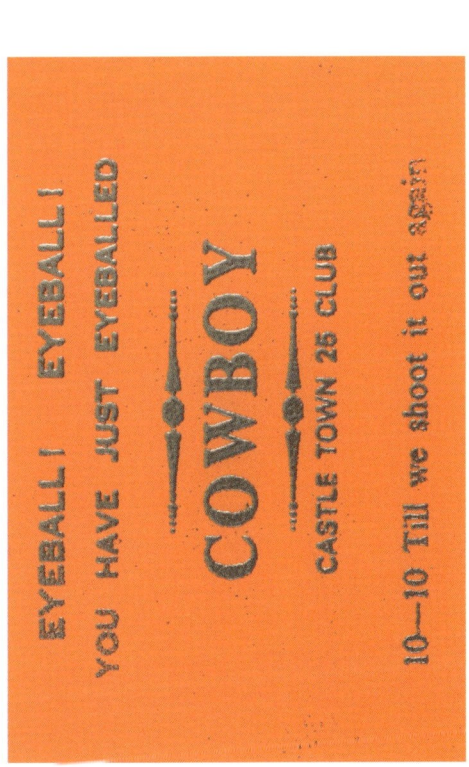

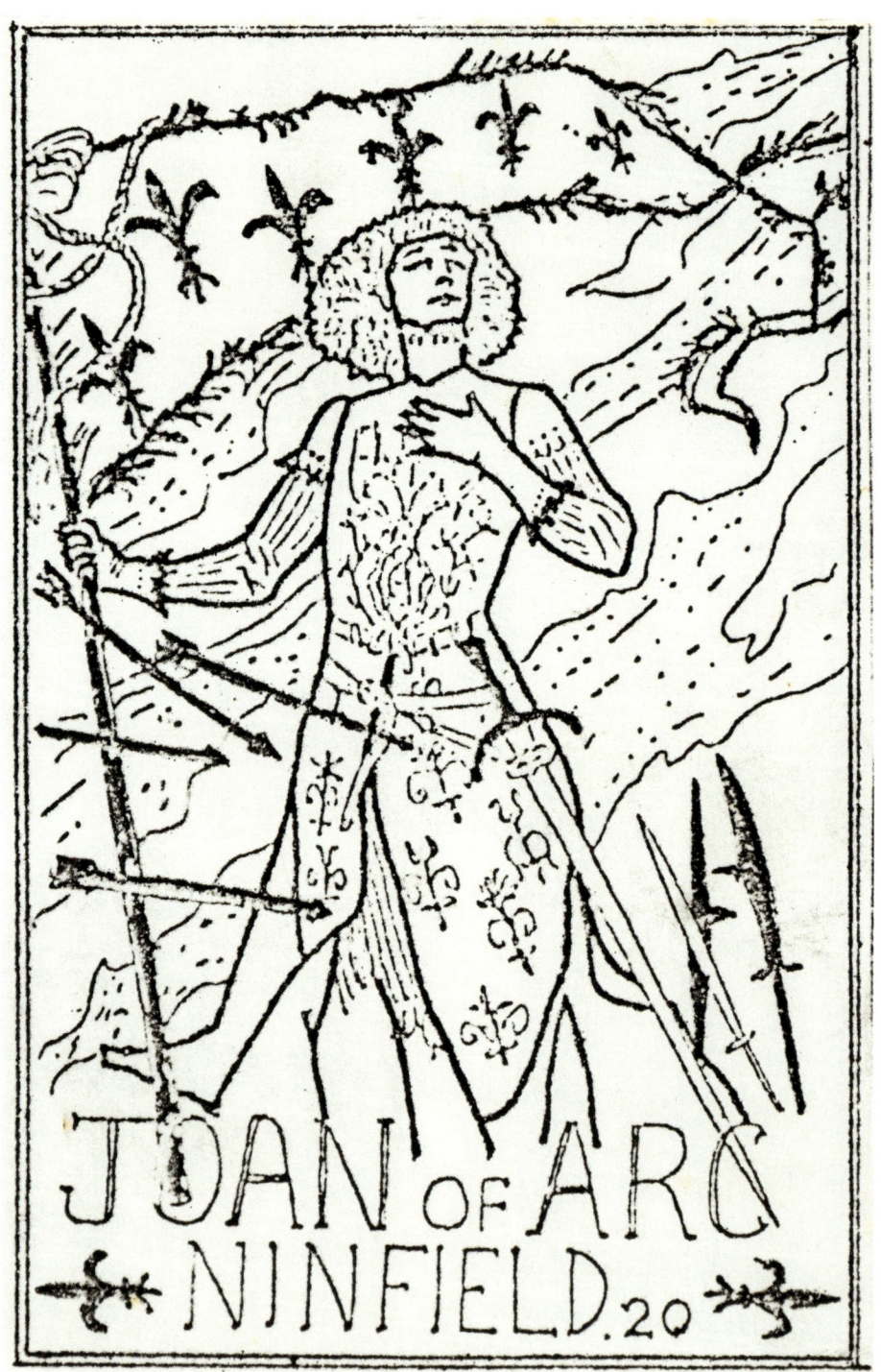

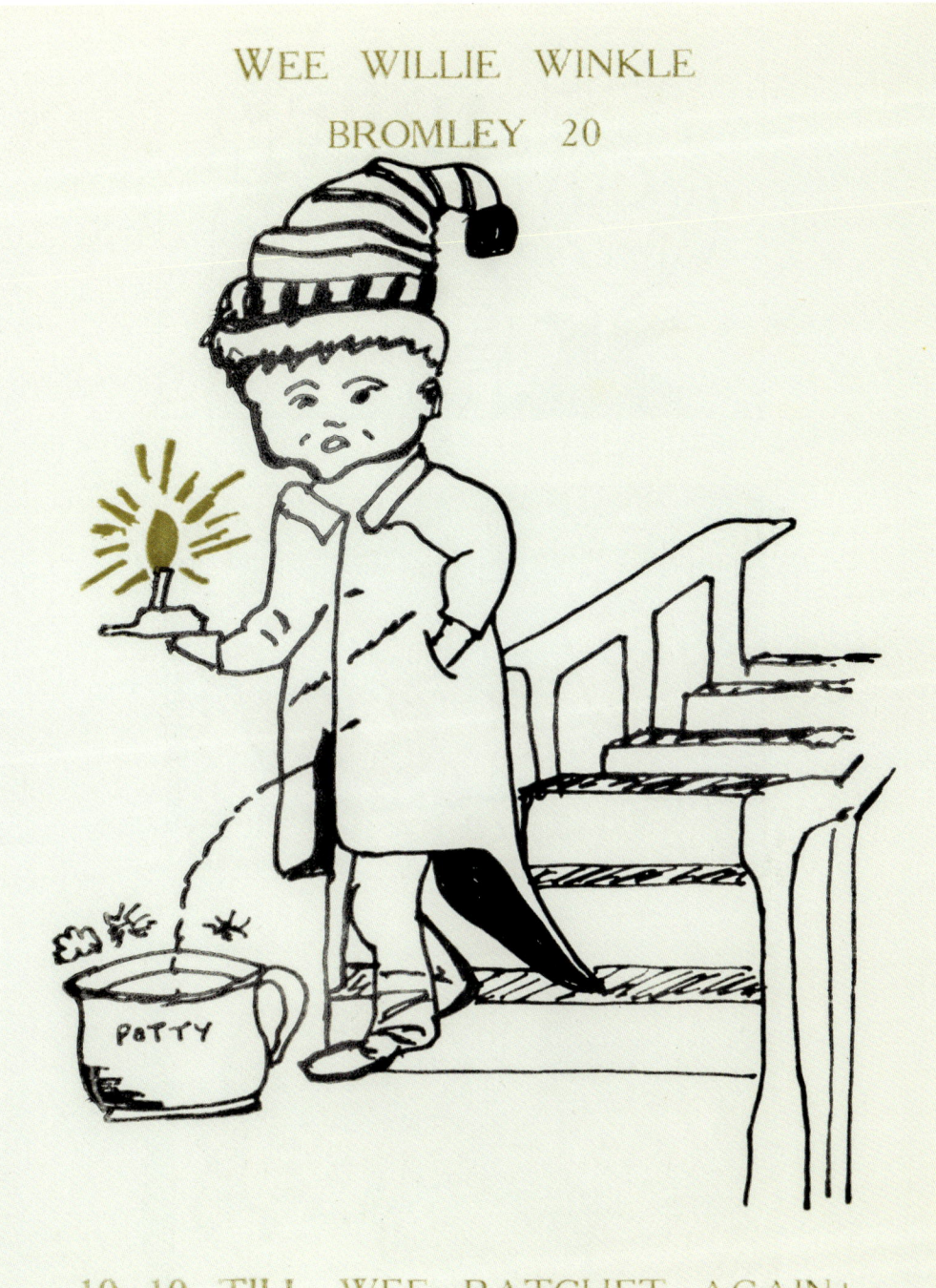

10-10 TILL WEE RATCHET AGAIN!

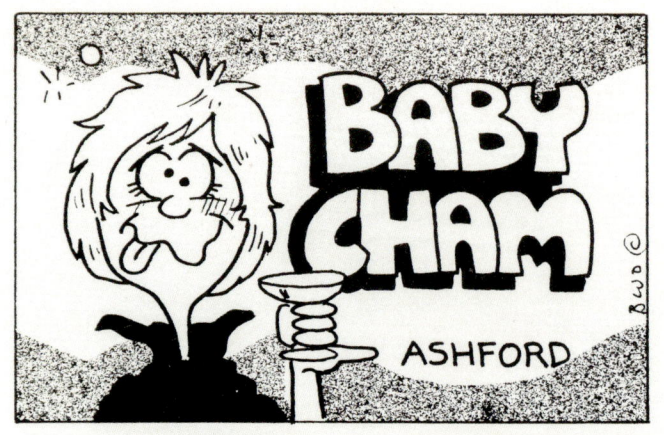

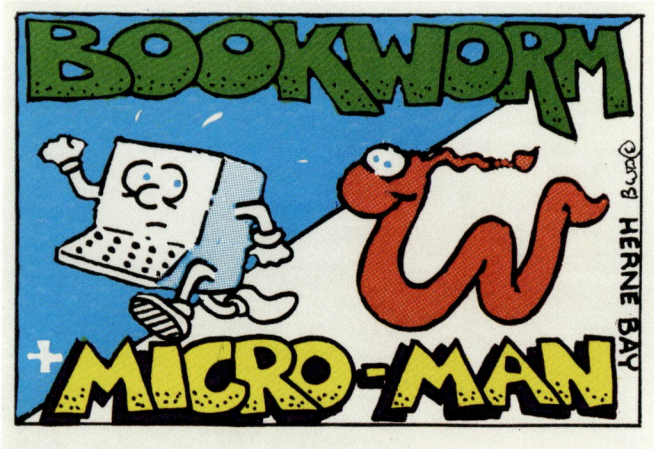

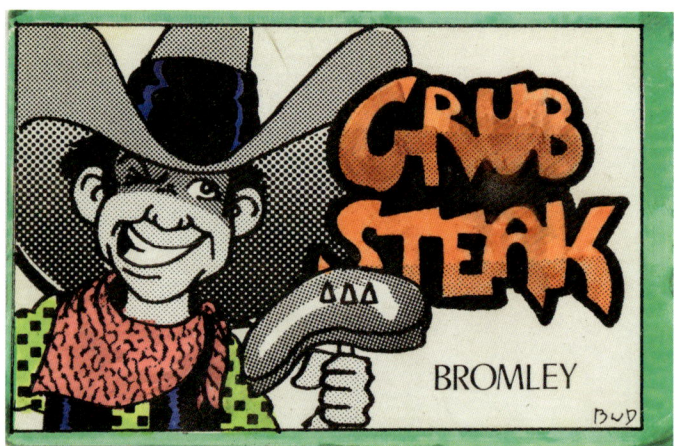

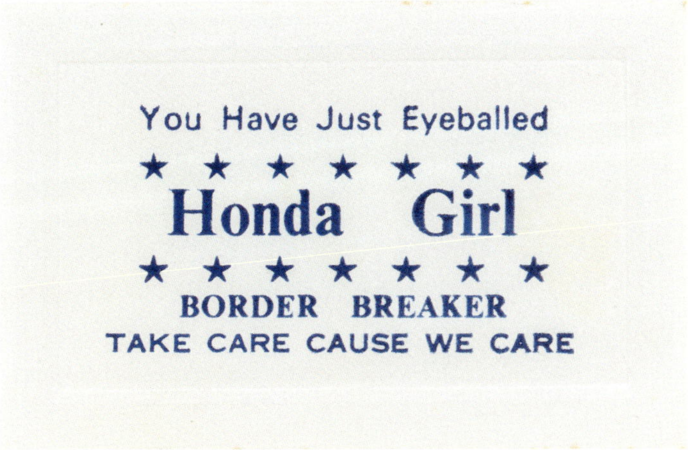

You Have Just Eyeballed

★ ★ ★ ★ ★ ★ ★ ★

Honda Girl

★ ★ ★ ★ ★ ★ ★ ★

BORDER BREAKER
TAKE CARE CAUSE WE CARE

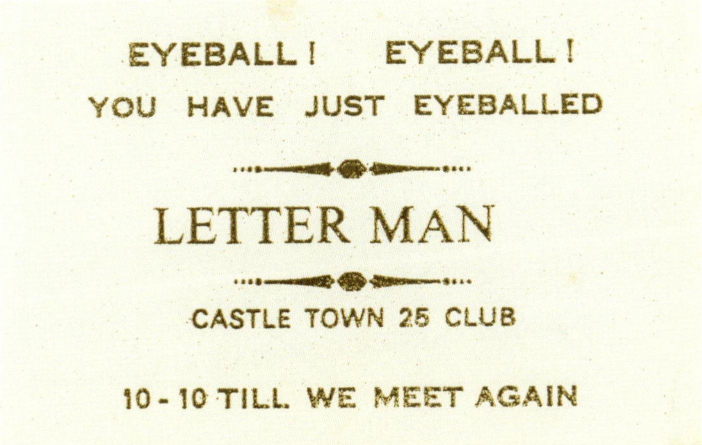

EYEBALL !   EYEBALL !
YOU HAVE JUST EYEBALLED

LETTER MAN

CASTLE TOWN 25 CLUB

10 - 10 TILL WE MEET AGAIN

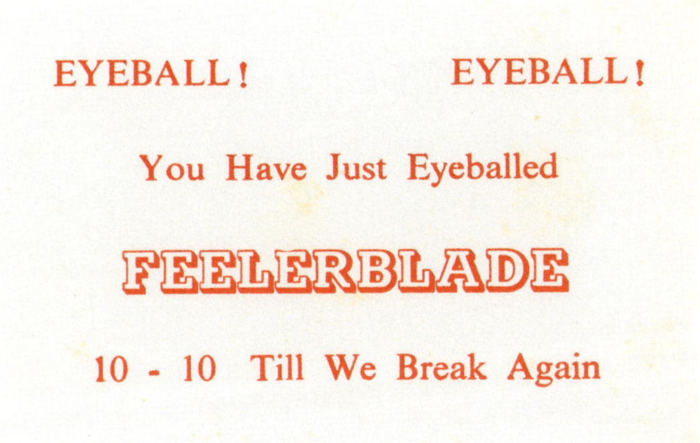

EYEBALL !   EYEBALL !

You Have Just Eyeballed

FEELERBLADE

10 - 10 Till We Break Again

Congratulations
you have just eyeballed

" AIR VENT "

Bergholt Breaker
take care 'cos I care

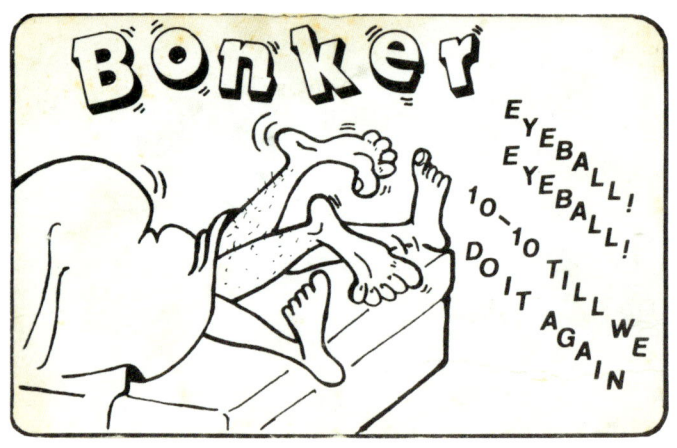

Eyeball   Eyeball
You've just eyeballed!!
88s ASH TRAY 73s
1-40 don't be naughty
10:10 till we do it again

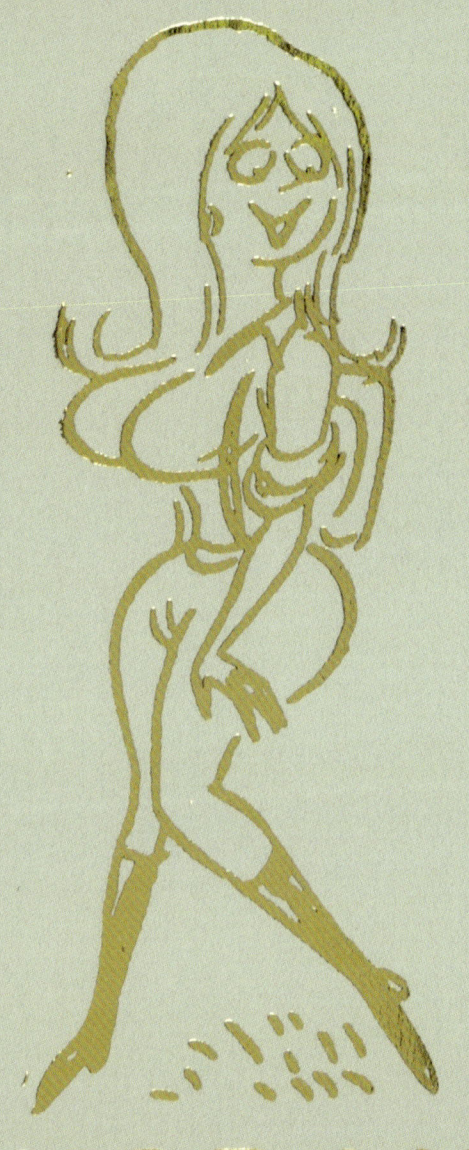

KODAK
LADY

EASTBOURNE 20

EYEBALL !

You Have J

Sweet

An

Doo

10-10 Till We

Rock On V

EYEBALL!

ust Eyeballed

Martini

d

bie

Break Again

We're Gone

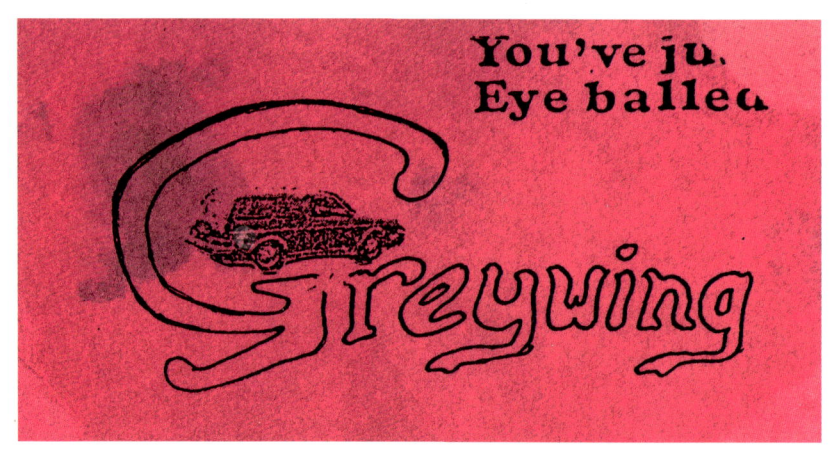

You've ju.
Eye balled

Greywing

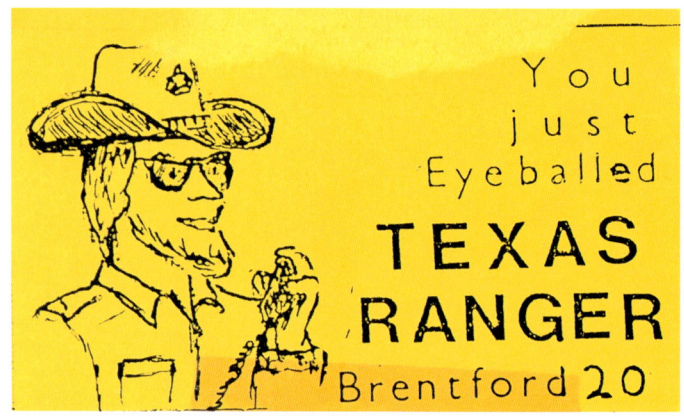

You
just
Eyeballed

TEXAS
RANGER

Brentford 20

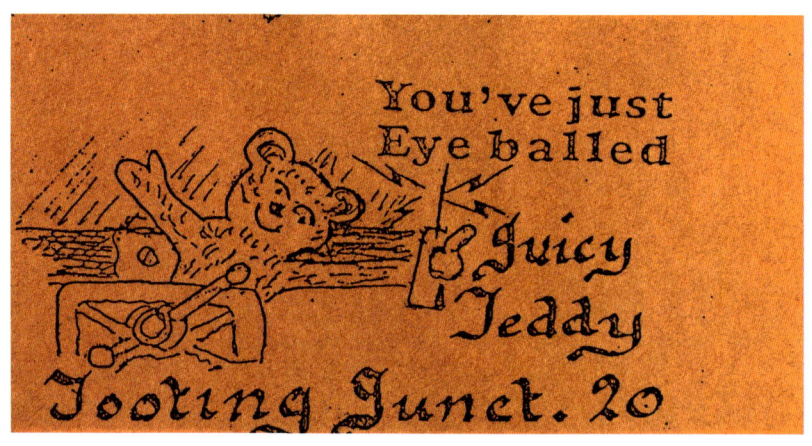

You've just
Eye balled

Juicy
Jeddy

Jooting Junct. 20

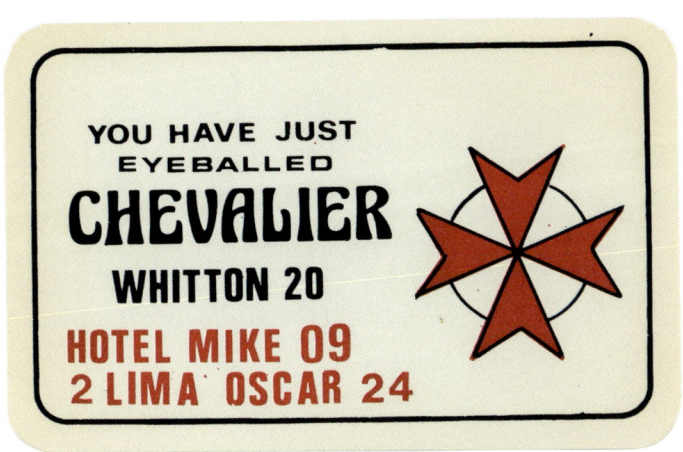

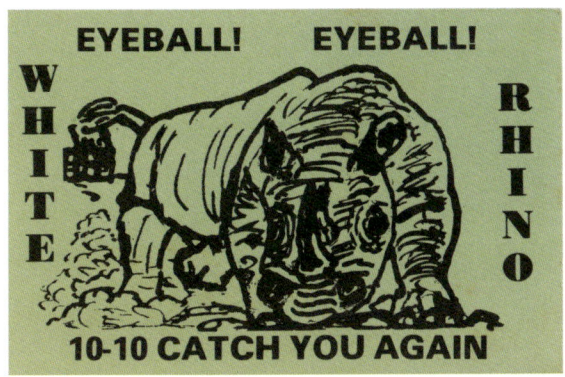

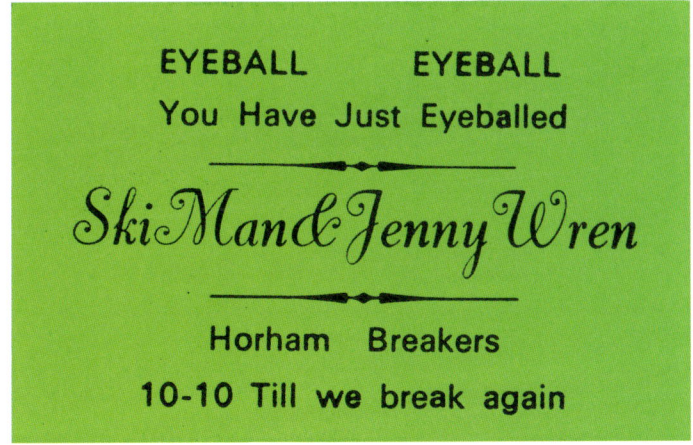

You Have Just Eyeballed

INTERNATIONAL
MAN

From The Burrelton 20

Take Care Cos' I Care

EYEBALL !   EYEBALL !

You Have Just Eyeballed

****

Disco Roller

****

Border   Breakers

10 – 10 Till We Roll Again

You have just had an Eye-ball
with

Goodwin

from   Sausage CIty

10   10   Breaker Break

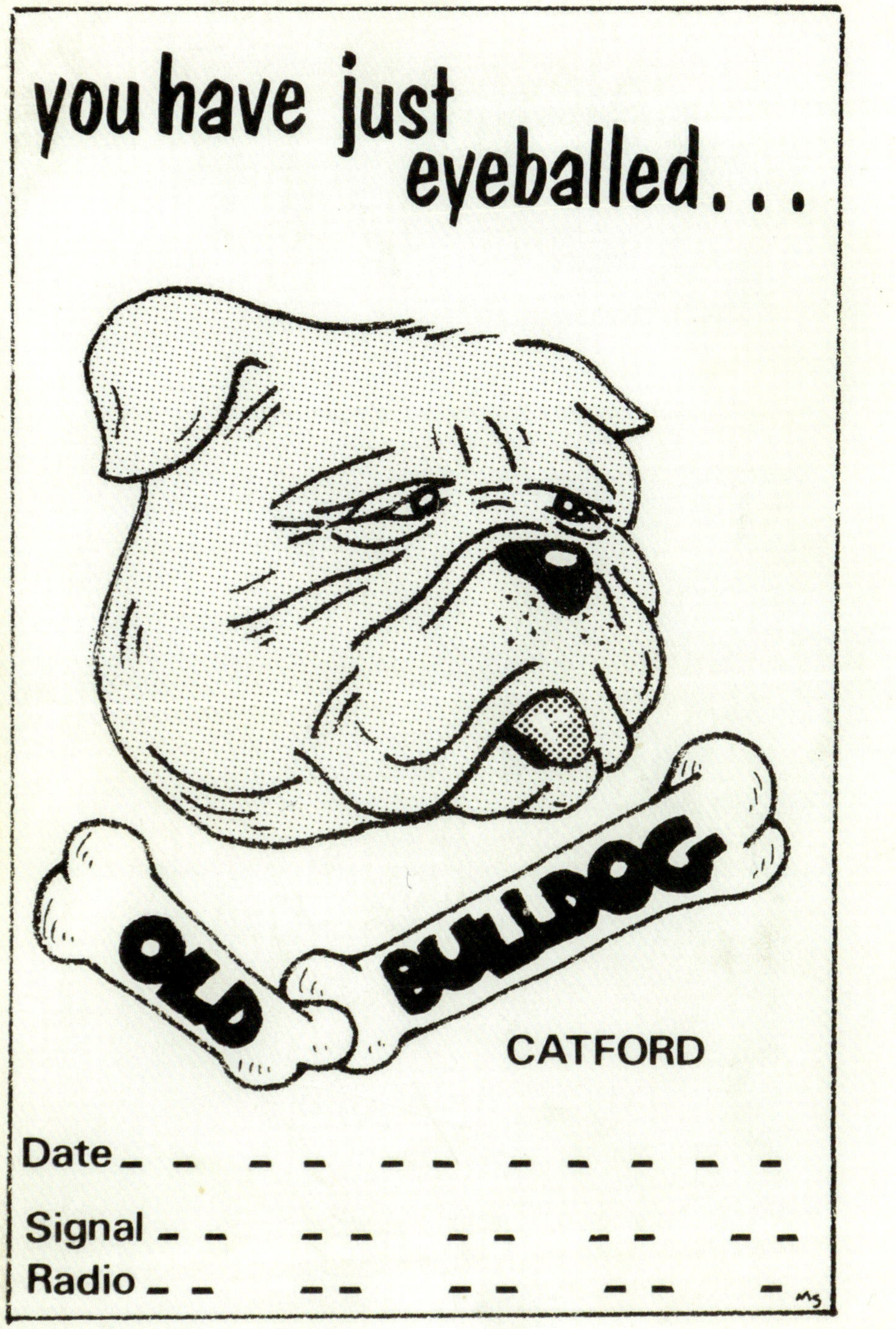

you have just eyeballed...

OLD BULLDOG

CATFORD

Date _ _ _ _ _ _ _ _ _ _

Signal _ _ _ _ _ _ _ _ _

Radio _ _ _ _ _ _ _ _ _ _

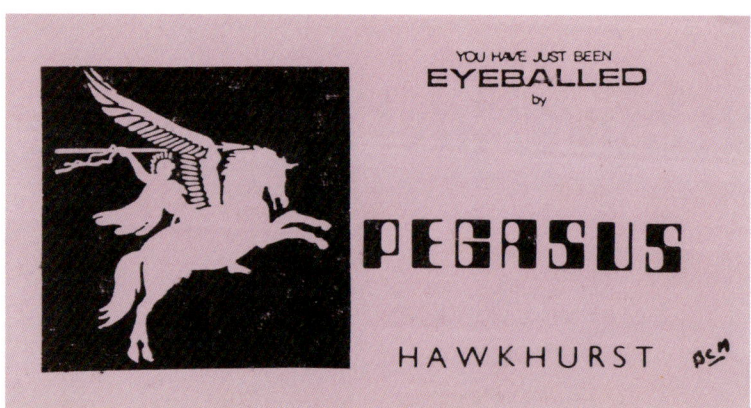

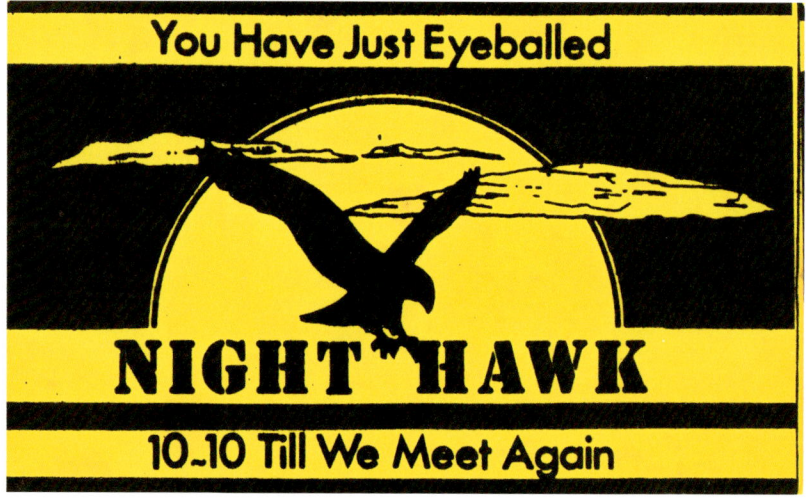

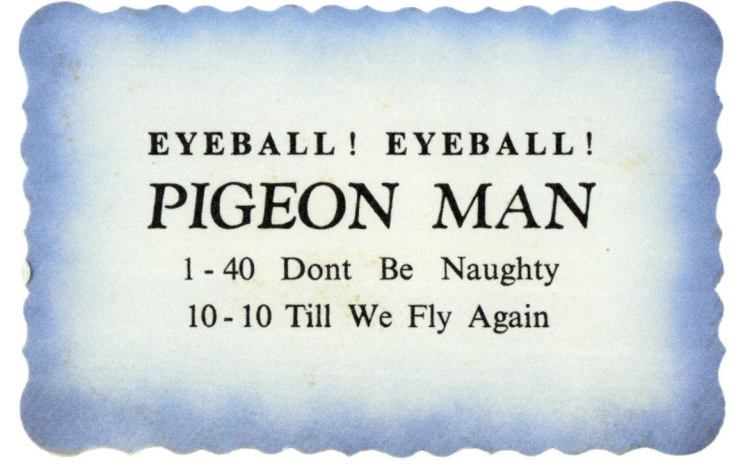

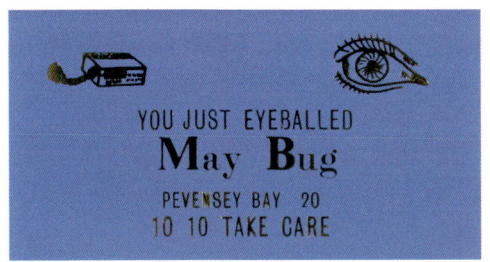

YOU JUST EYEBALLED
May Bug
PEVENSEY BAY 20
10 10 TAKE CARE

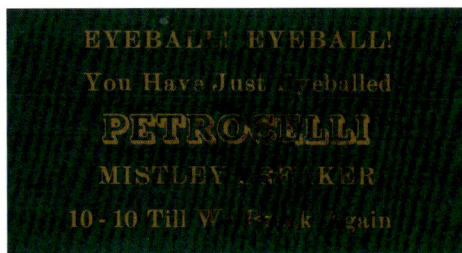

EYEBALL! EYEBALL!
You Have Just Eyeballed
PETROSELLI
MISTLEY BREAKER
10 - 10 Till We Break Again

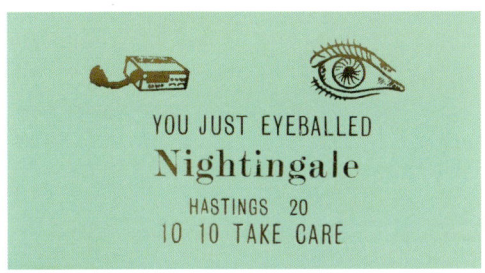

YOU JUST EYEBALLED
Nightingale
HASTINGS 20
10 10 TAKE CARE

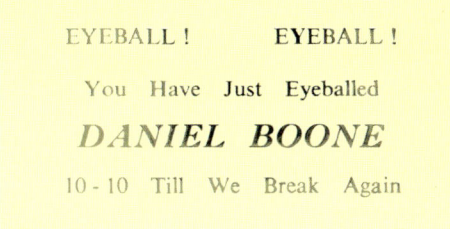

EYEBALL !   EYEBALL !
You Have Just Eyeballed
DANIEL BOONE
10 - 10 Till We Break Again

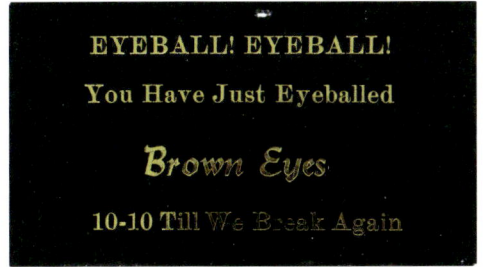

EYEBALL! EYEBALL!
You Have Just Eyeballed
Brown Eyes
10-10 Till We Break Again

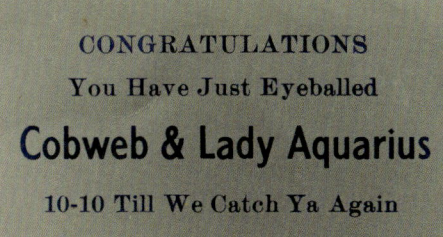

CONGRATULATIONS
You Have Just Eyeballed
Cobweb & Lady Aquarius
10-10 Till We Catch Ya Again

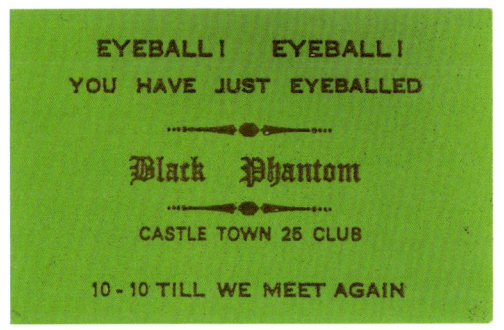

EYEBALL !   EYEBALL !
YOU HAVE JUST EYEBALLED
Black   Phantom
CASTLE TOWN 25 CLUB
10 - 10 TILL WE MEET AGAIN

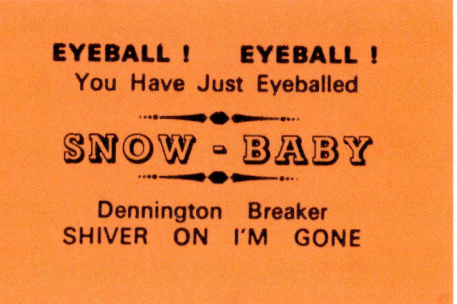

EYEBALL !   EYEBALL !
You Have Just Eyeballed
SNOW - BABY
Dennington Breaker
SHIVER ON I'M GONE

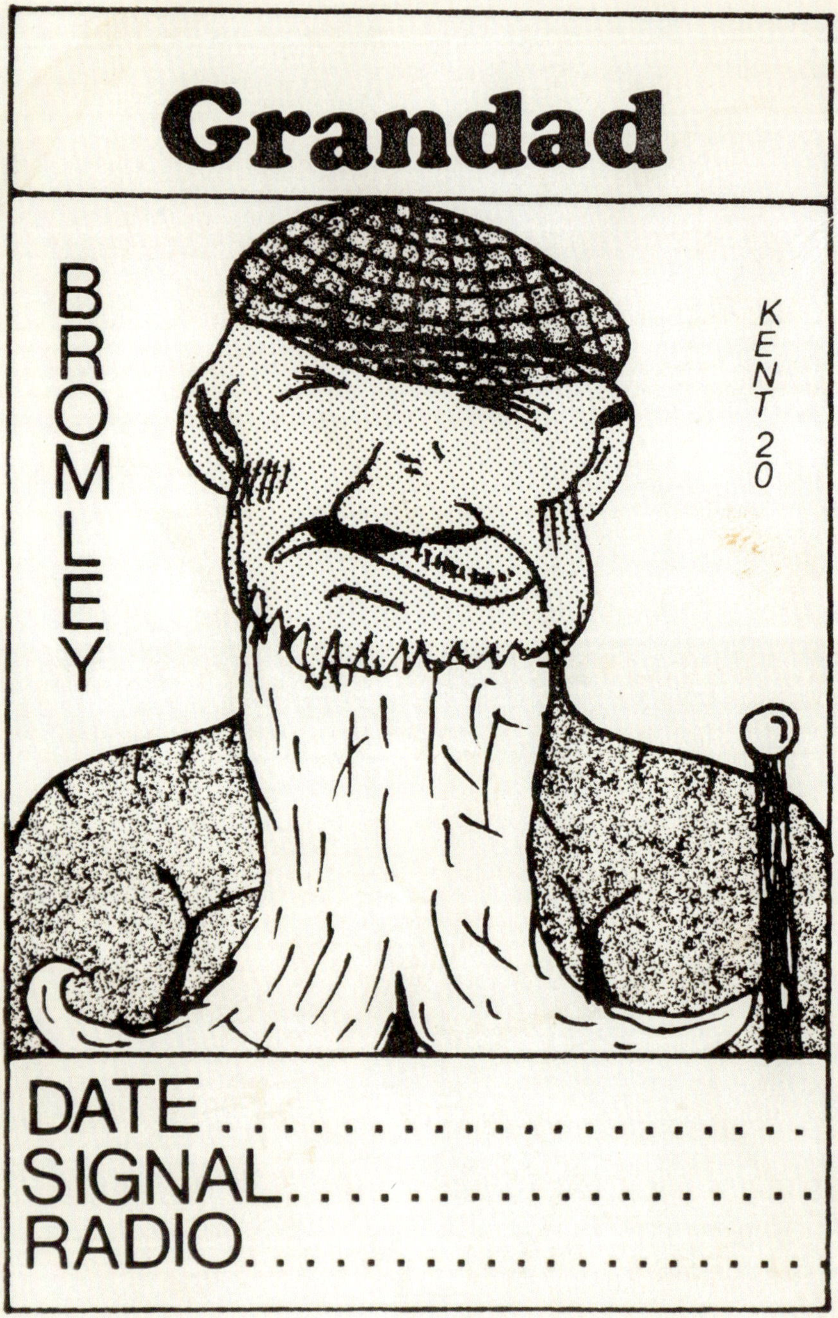

Grandad

BROMLEY

KENT 20

DATE . . . .
SIGNAL . . . .
RADIO . . . .

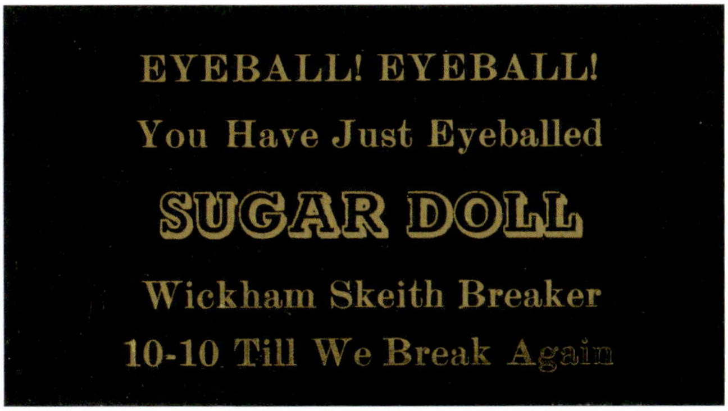

EYEBALL! EYEBALL!

You Have Just Eyeballed

SUGAR DOLL

Wickham Skeith Breaker

10-10 Till We Break Again

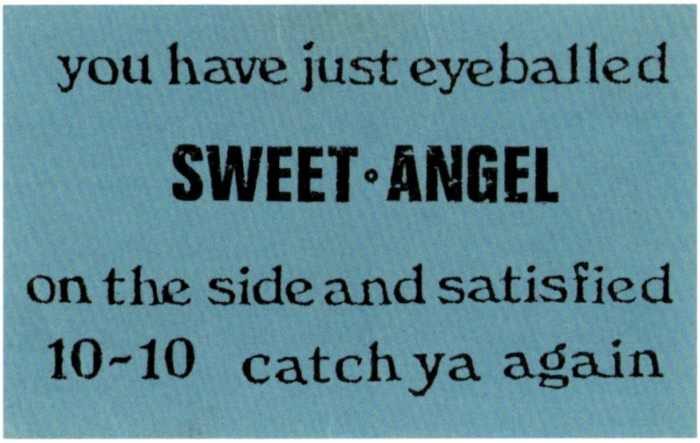

you have just eyeballed

SWEET·ANGEL

on the side and satisfied

10~10 catch ya again

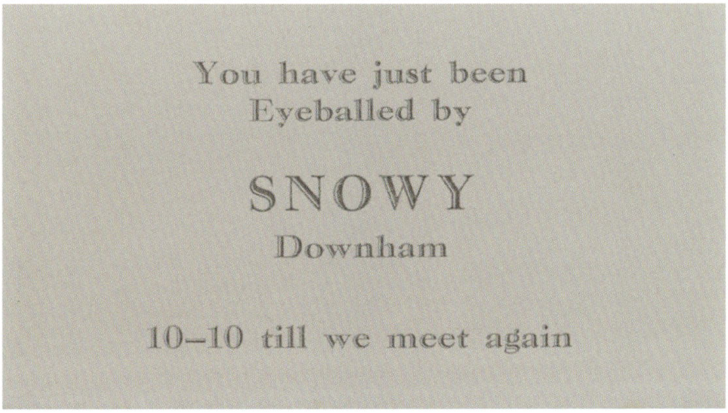

You have just been
Eyeballed by

SNOWY

Downham

10–10 till we meet again

YOU HAVE JUST E

Radar &

Gil &

Unit 09   P. O

Ammanford, Dyf

ALL THE BEST TO

EBALLED

Hatpin

Pat

Box 5

d. S A 18 3 BN

YOU & YOURS

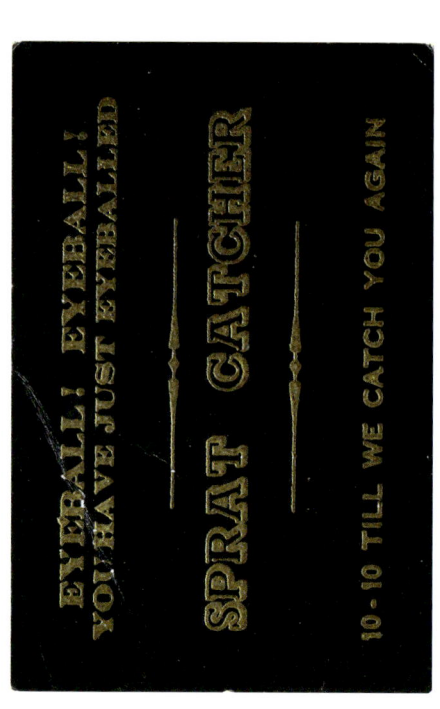

EYEBALL! EYEBALL!
YOU HAVE JUST EYEBALLED

SPRAT CATCHER

10-10 TILL WE CATCH YOU AGAIN

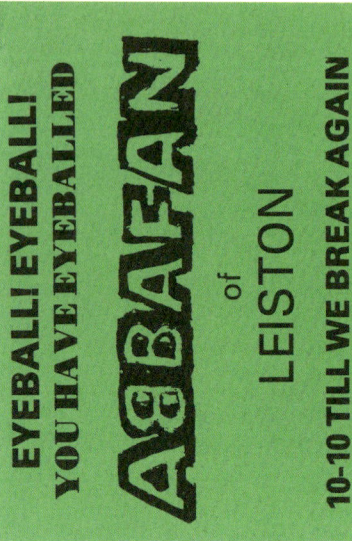

EYEBALL! EYEBALL!
YOU HAVE EYEBALLED

ABBAFAN

of

LEISTON

10-10 TILL WE BREAK AGAIN

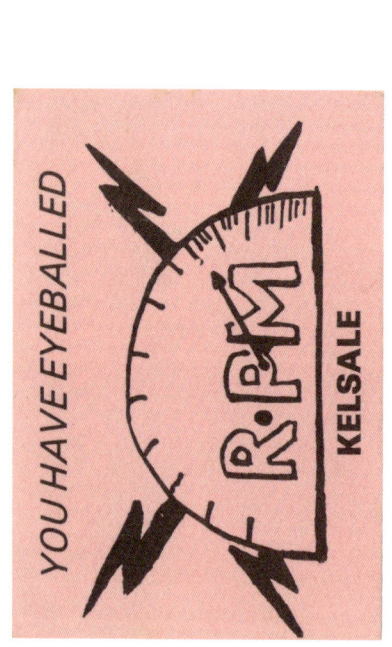

YOU HAVE EYEBALLED

R.P.M

KELSALE

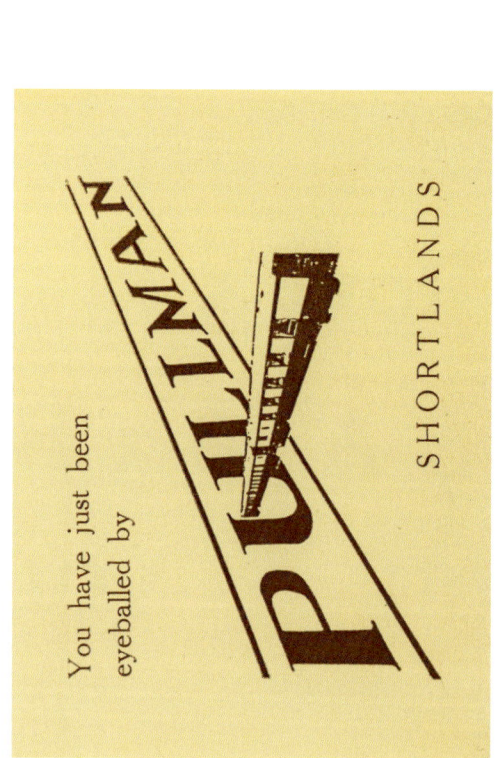

You have just been eyeballed by

PULLMAN

SHORTLANDS

You Have Eyeballed

Grandad

Roy

Bexhill   20

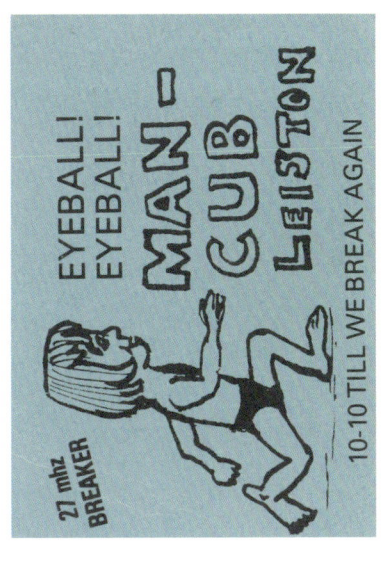

27 mhz
BREAKER

EYEBALL!
EYEBALL!

MAN-
CUB
LEISTON

10-10 TILL WE BREAK AGAIN

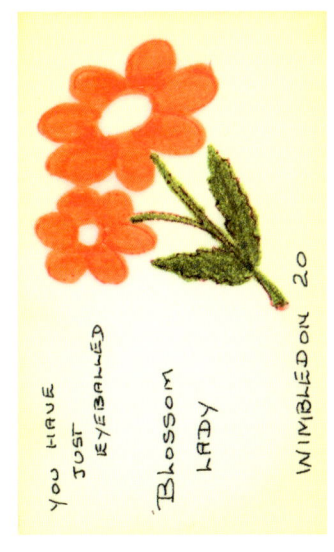

YOU HAVE
JUST
EYEBALLED

'Blossom
LADY

WIMBLEDON 20

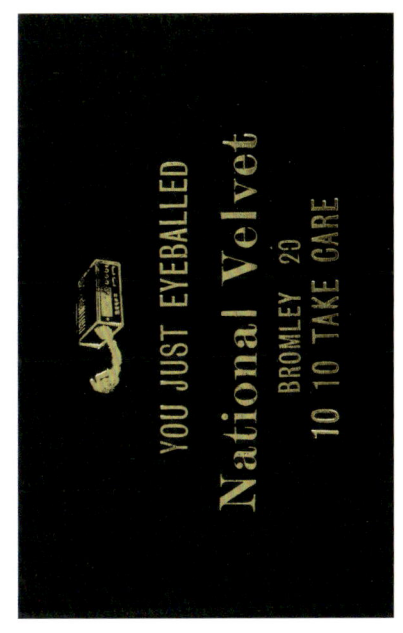

YOU JUST EYEBALLED

National Velvet

BROMLEY 20

10 10 TAKE CARE

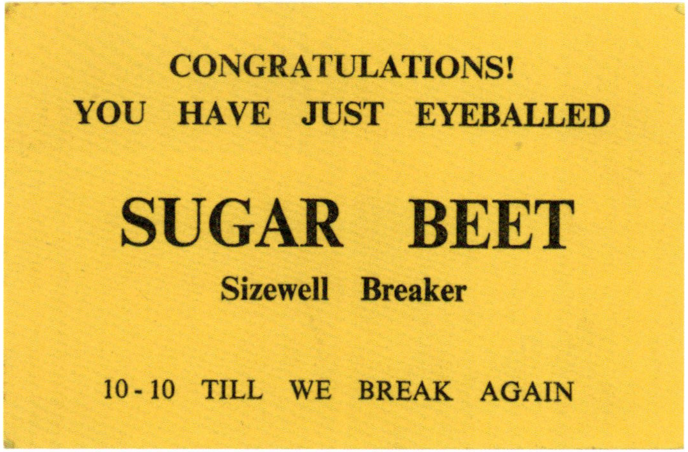

CONGRATULATIONS!
YOU HAVE JUST EYEBALLED

SUGAR BEET

Sizewell Breaker

10-10 TILL WE BREAK AGAIN

Bob Dye, aka 'Sugar Beet'.

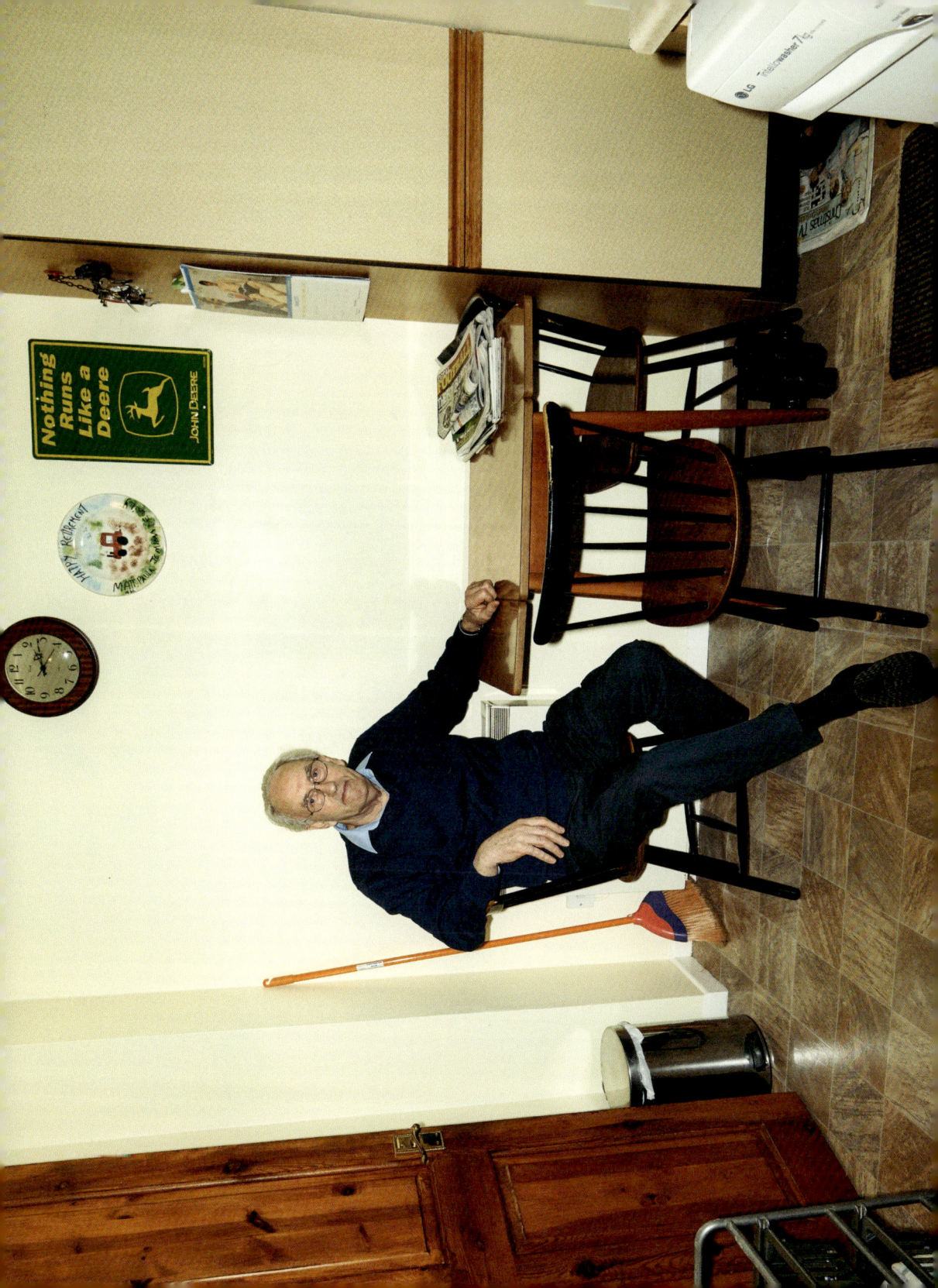

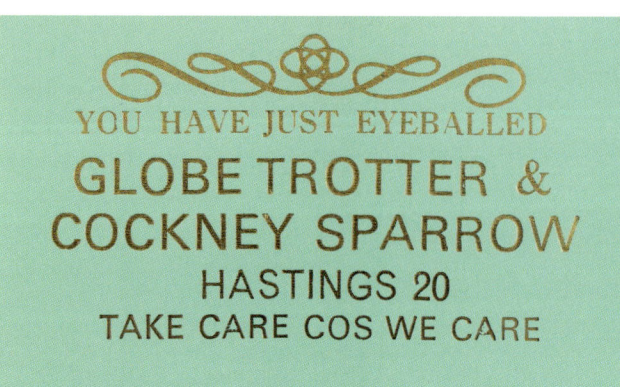

YOU HAVE JUST EYEBALLED
GLOBE TROTTER &
COCKNEY SPARROW
HASTINGS 20
TAKE CARE COS WE CARE

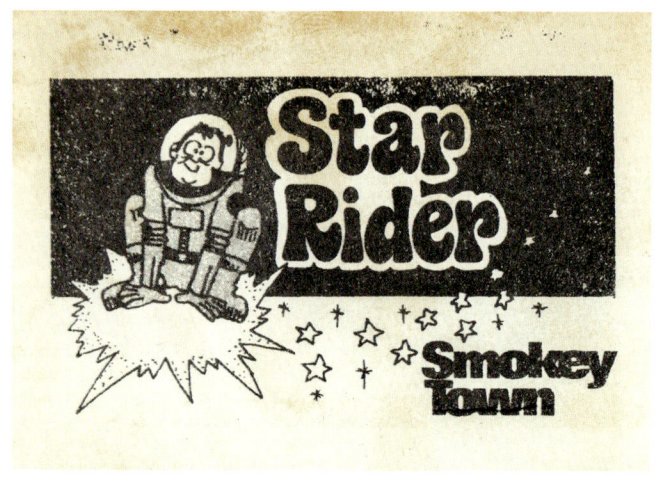

You've Just Eyeballed

IRONSIDES

Q.T.H   Bedfont   F/P ALF

YOU HAVE JUST EYEBALLED

W   T
 I   A   A
  M   L   L
   B   L   * GIRL
    L   E
     2   D   R
      D   O   L
           N

IO-IO   TILL WE MEET AGAIN

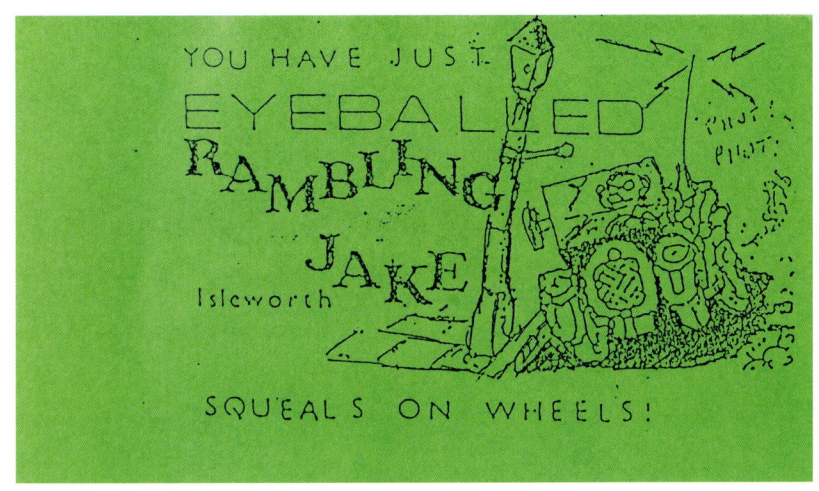

YOU HAVE JUST
EYEBALLED
RAMBLING
JAKE
Isleworth
PILOT
PILOT
SQUEALS ON WHEELS!

EYEBALL! EYEBALL!
You Have Just Eyeballed
DIRT DEMON
MY Back Wheels Spinnen
And I'm Still Grinnen
10 - 10

73's Persona

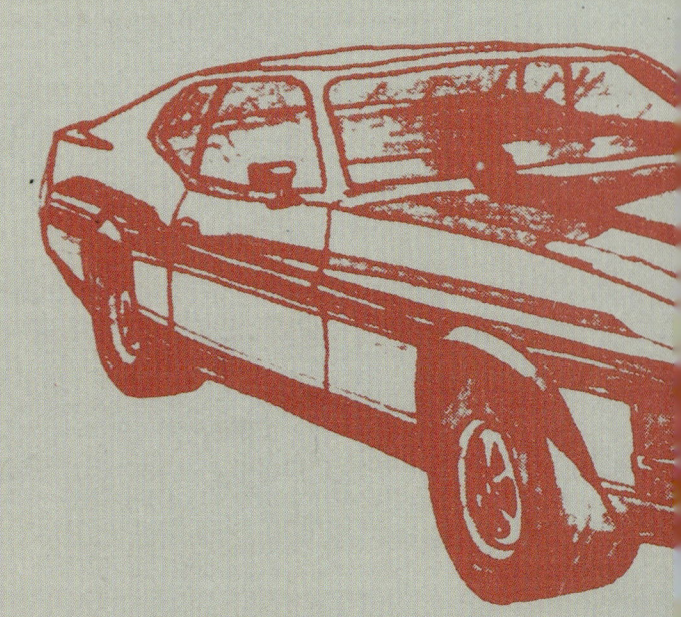

HEAVY

P.O. BOX 100, COXHEA

: NICK 51's

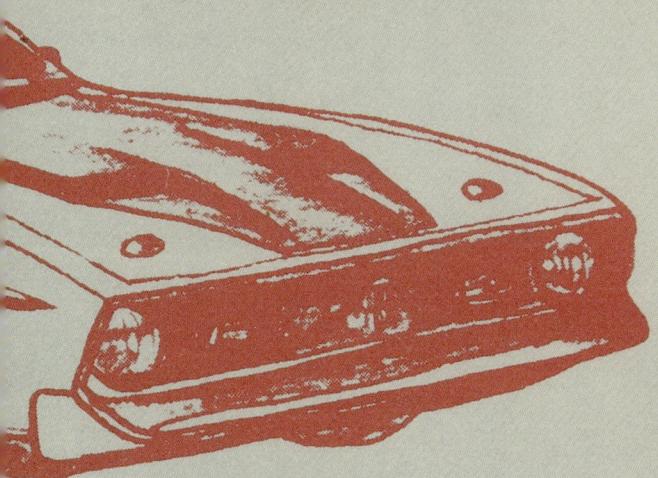

METAL

TH, MAIDSTONE, KENT

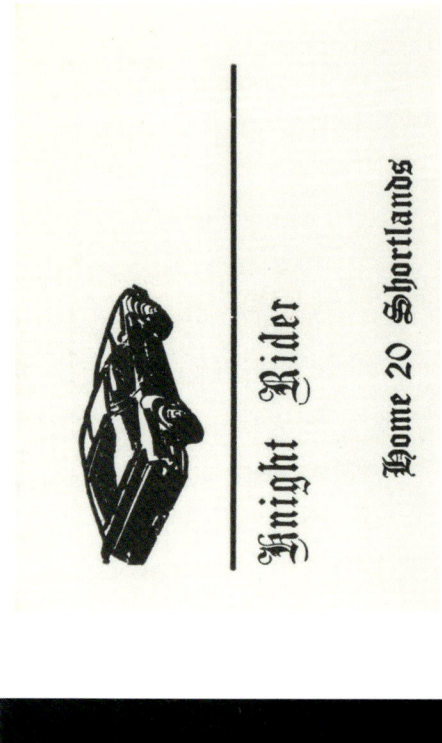

Knight Rider

Home 20 Shortlands

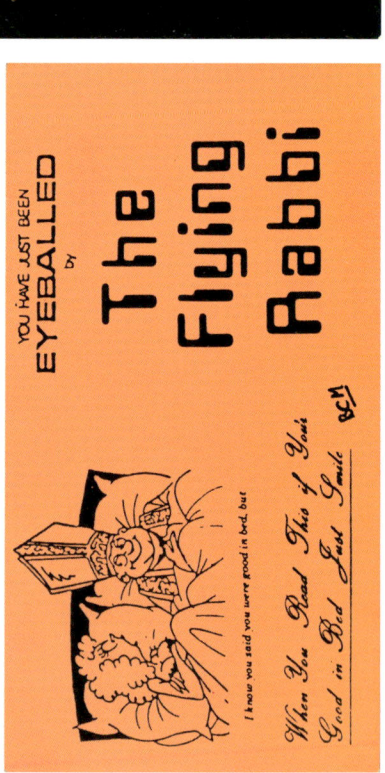

EYEBALL! EYEBALL!

You Have Just Eyeballed

NIGHT RIDER

10-10 Till We Break Again

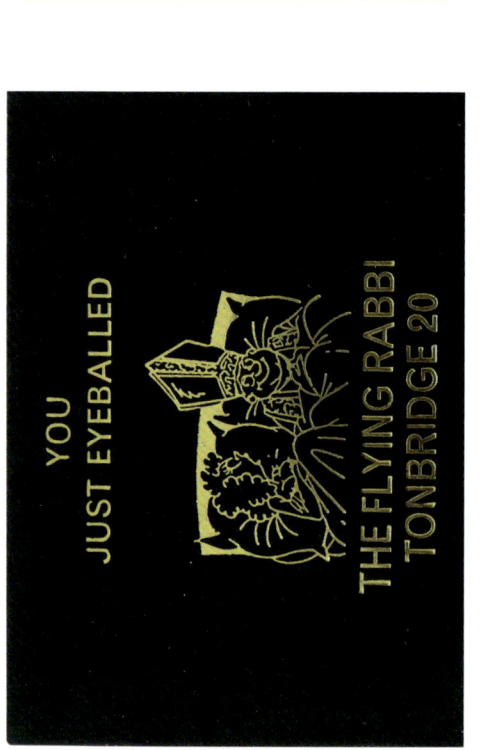

YOU
JUST EYEBALLED

THE FLYING RABBI
TONBRIDGE 20

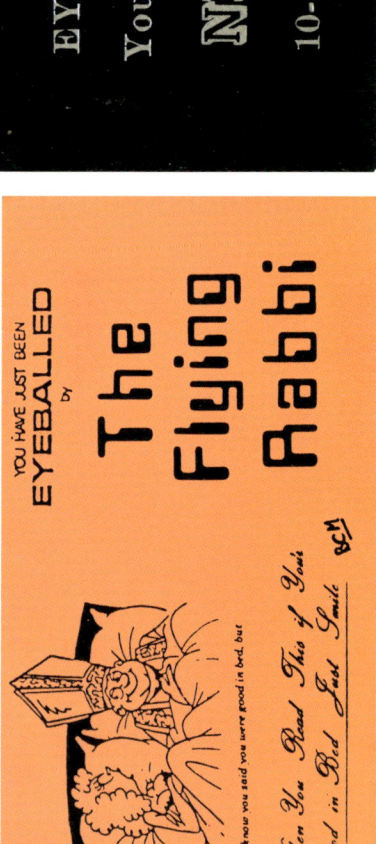

YOU HAVE JUST BEEN
EYEBALLED
by

The
Flying
Rabbi

I know you said you were good in bed, but

When You Read This of Yours
Good in Bed Just Smile
BCM

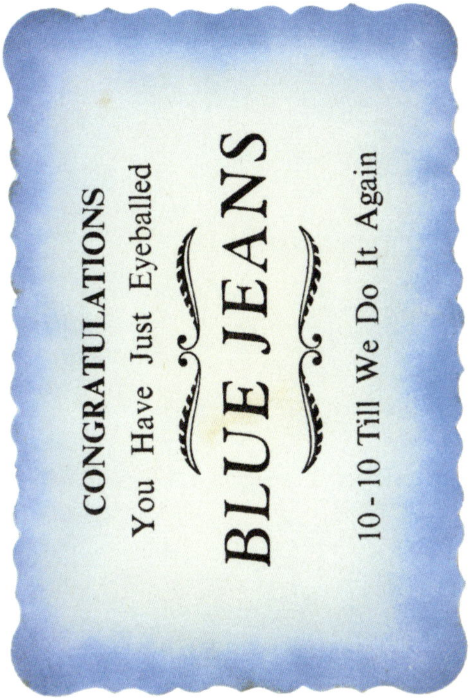

CONGRATULATIONS

You Have Just Eyeballed

BLUE JEANS

10 - 10 Till We Do It Again

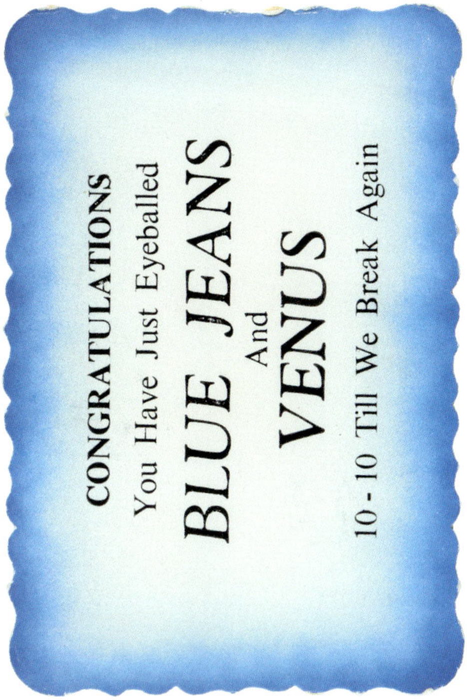

CONGRATULATIONS

You Have Just Eyeballed

BLUE JEANS

And

VENUS

10 - 10 Till We Break Again

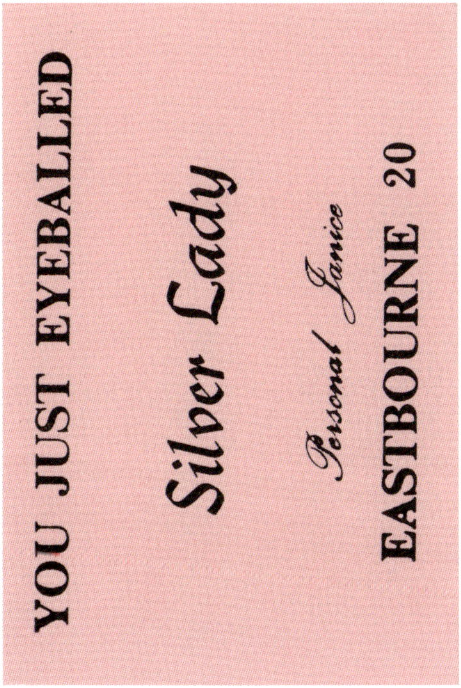

YOU JUST EYEBALLED

Silver Lady

Personal Janice

EASTBOURNE 20

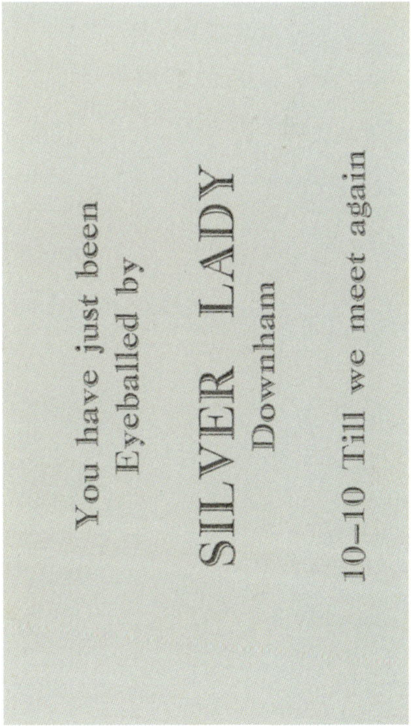

You have just been
Eyeballed by

SILVER LADY

Downham

10–10 Till we meet again

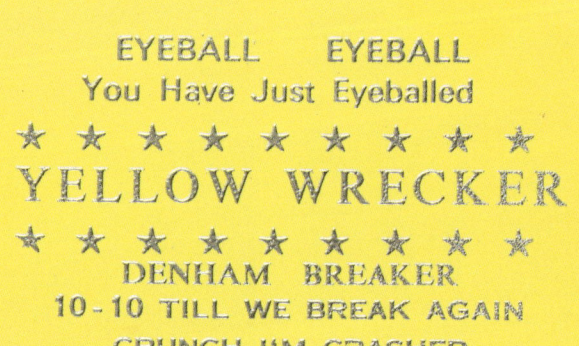

EYEBALL   EYEBALL
You Have Just Eyeballed

★ ★ ★ ★ ★ ★ ★ ★ ★
YELLOW WRECKER
★ ★ ★ ★ ★ ★ ★ ★ ★

DENHAM BREAKER
10-10 TILL WE BREAK AGAIN
CRUNCH I'M CRASHED

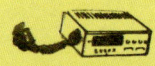

YOU JUST EYEBALLED

Ladybird
EDENBRIDGE   20
10 10 TAKE CARE

CONGRATULATIONS!
YOU HAVE JUST EYEBALLED

HUBBLE   **B**UBBLE

Bells Town Breaker

10-10 TILL WE BREAK AGAIN

Bitter Lemon

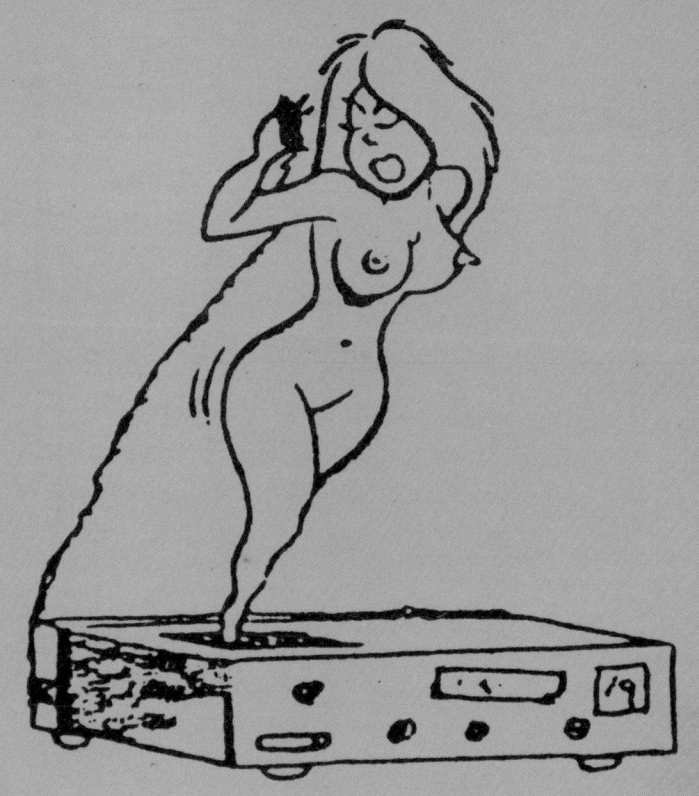

HASTINGS

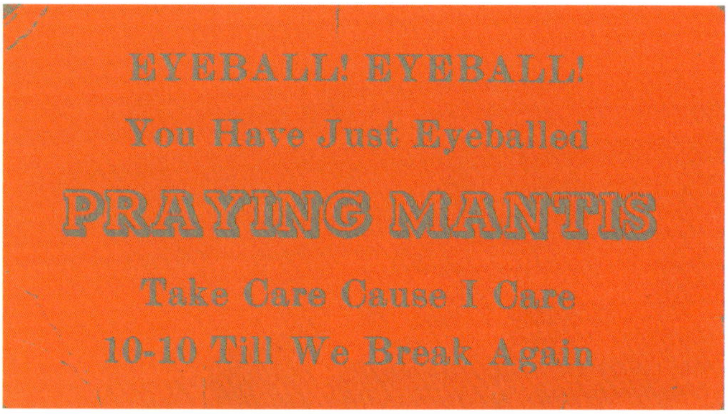

EYEBALL! EYEBALL!

You Have Just Eyeballed

PRAYING MANTIS

Take Care Cause I Care

10-10 Till We Break Again

Mark Alden, aka 'Praying Mantis'.

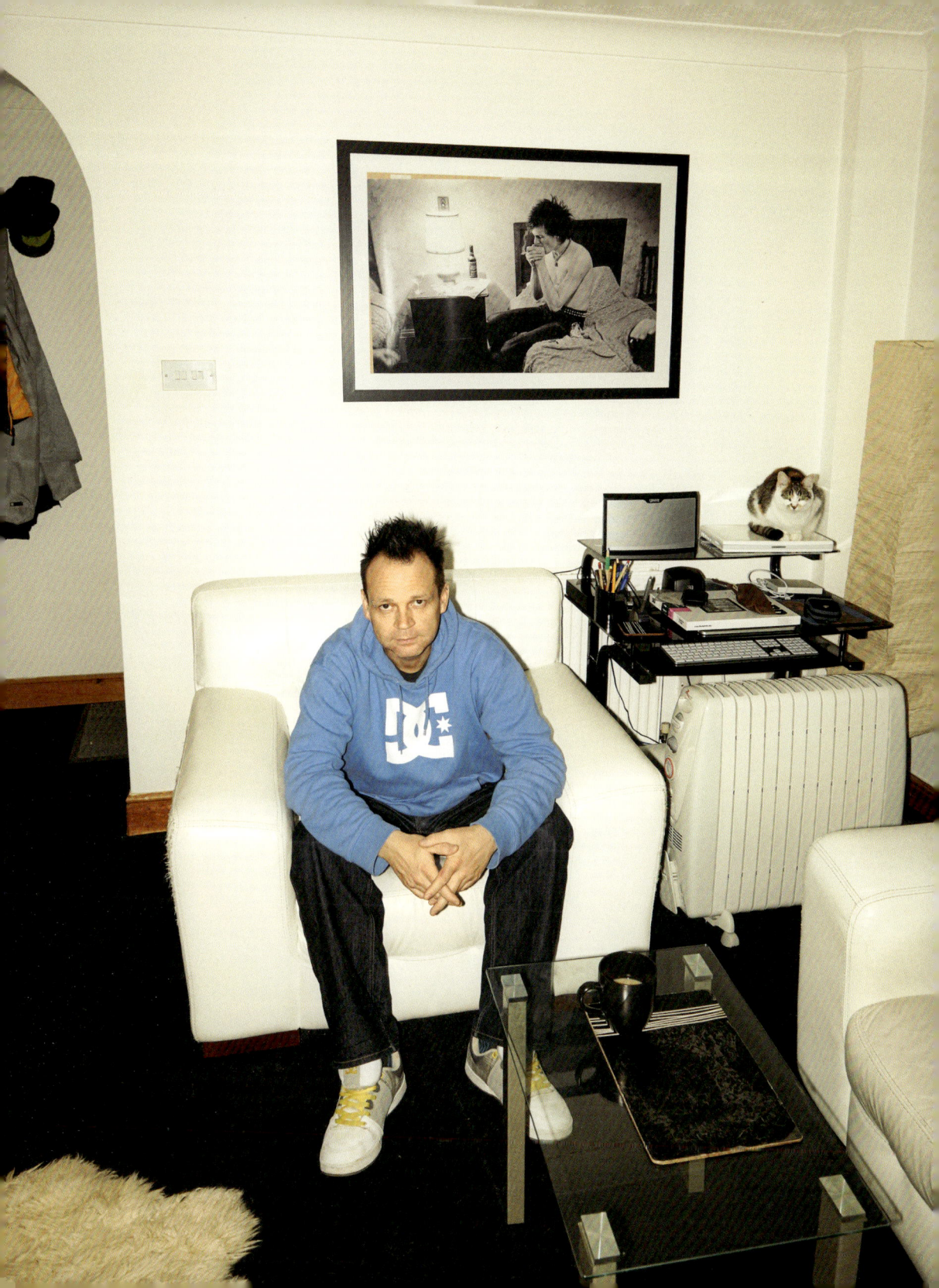

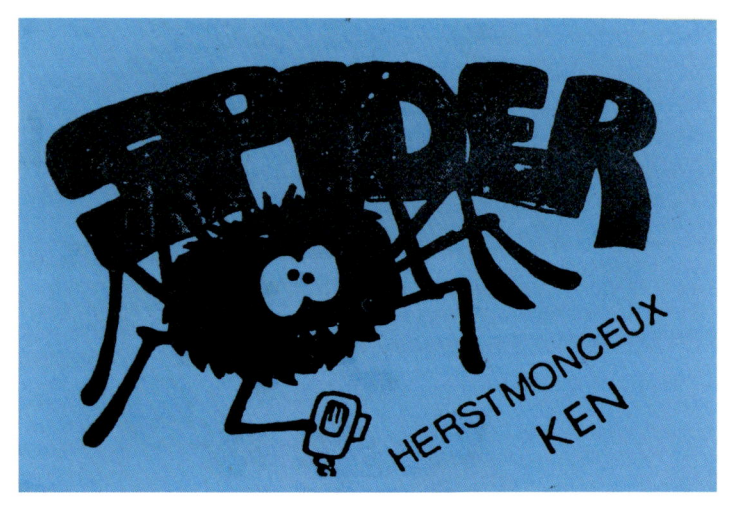

SPYDER THUNDER

HERSTMONCEUX KEN

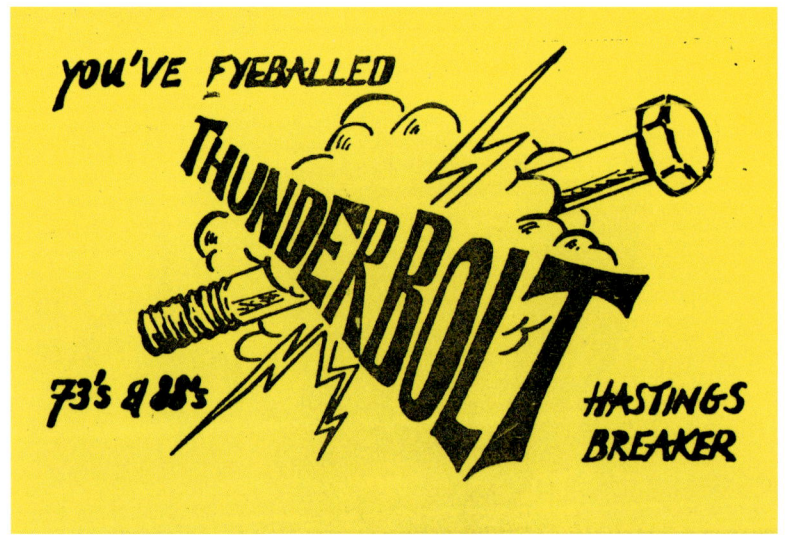

YOU'VE EYEBALLED

THUNDERBOLT

73's & 88's

HASTINGS BREAKER

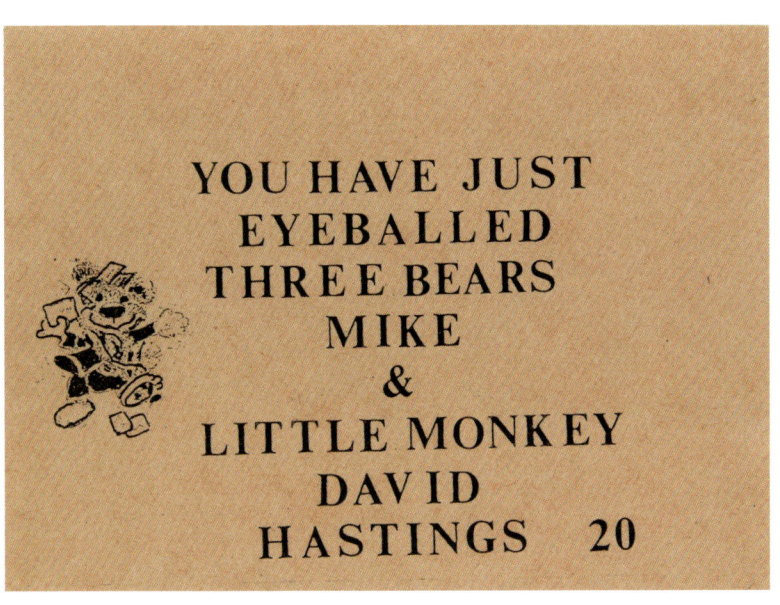

YOU HAVE JUST
EYEBALLED
THREE BEARS
MIKE
&
LITTLE MONKEY
DAVID
HASTINGS   20

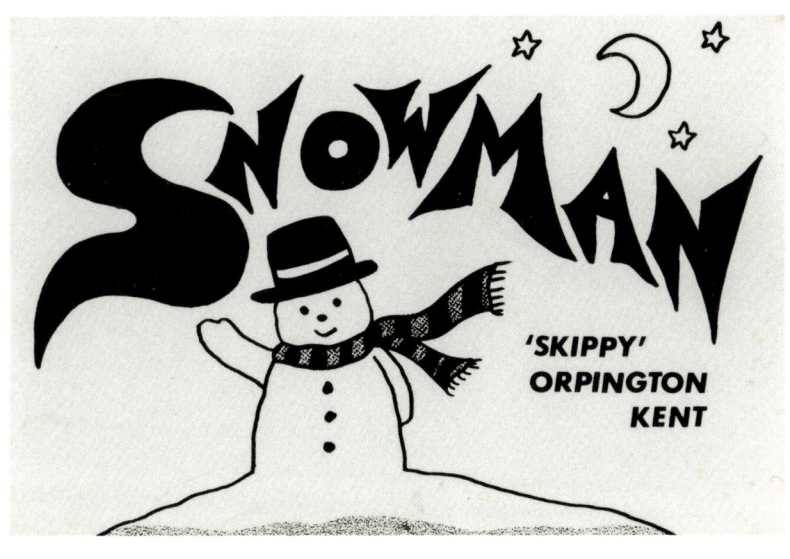

SNOWMAN

'SKIPPY'
ORPINGTON
KENT

YOU HAVE JUST B

BOX SP

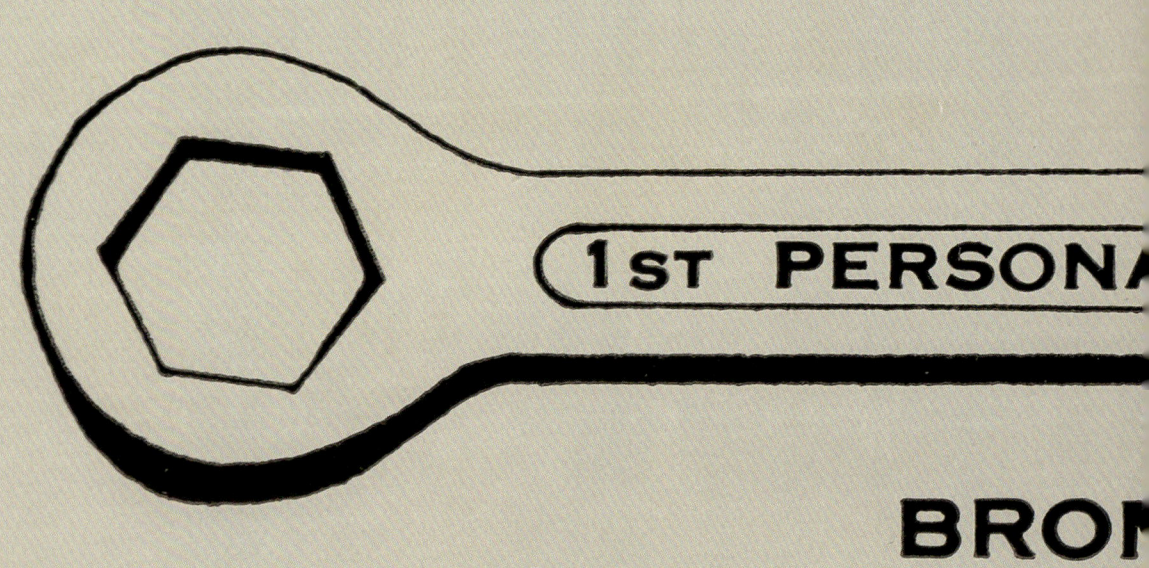

1st PERSONA

BRON

TAKE CARE BEC

EEN EYEBALLED BY

PANNER

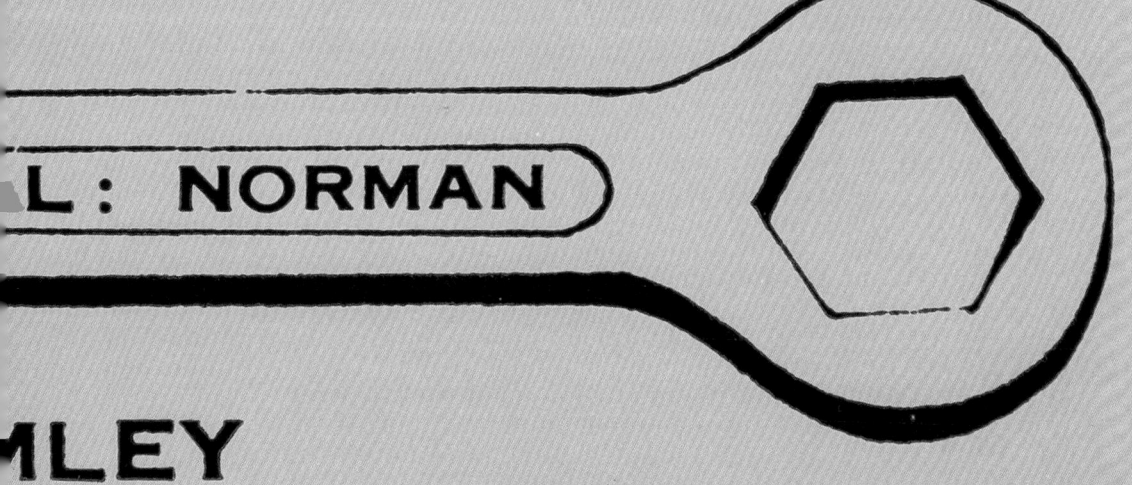

L: NORMAN

MLEY

AUSE WE CARE

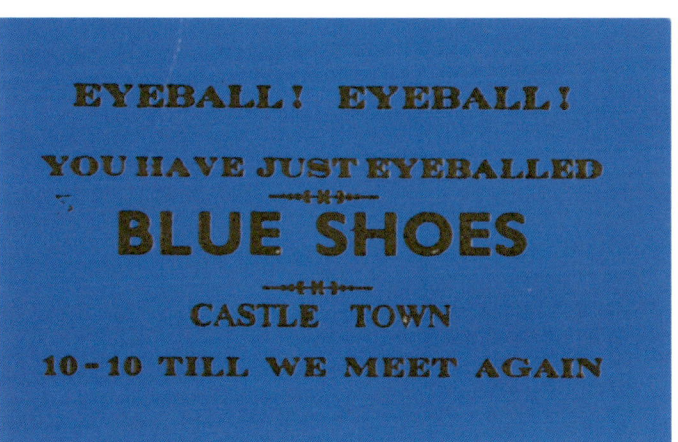

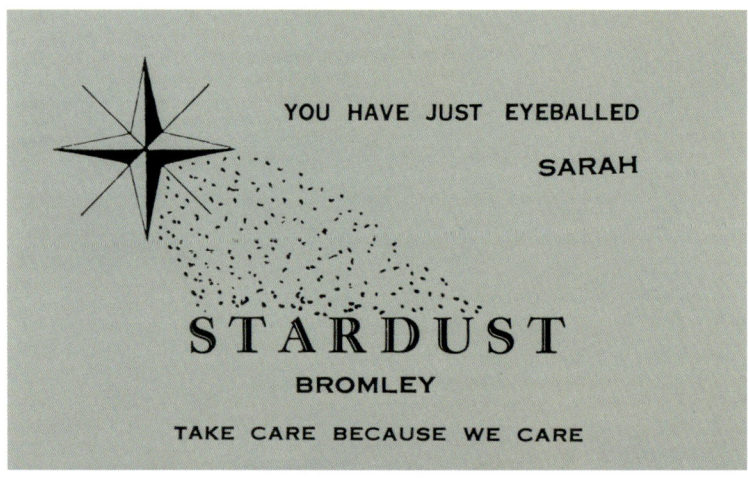

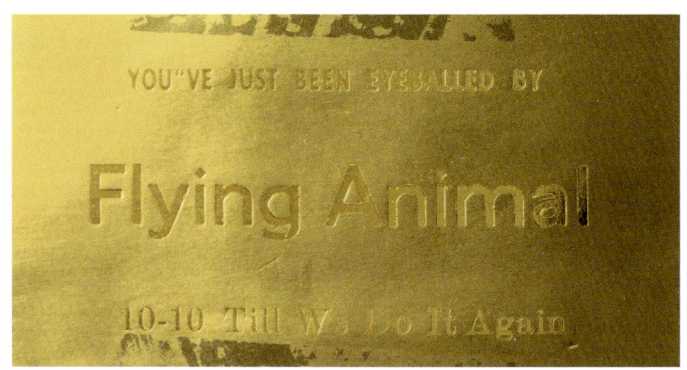

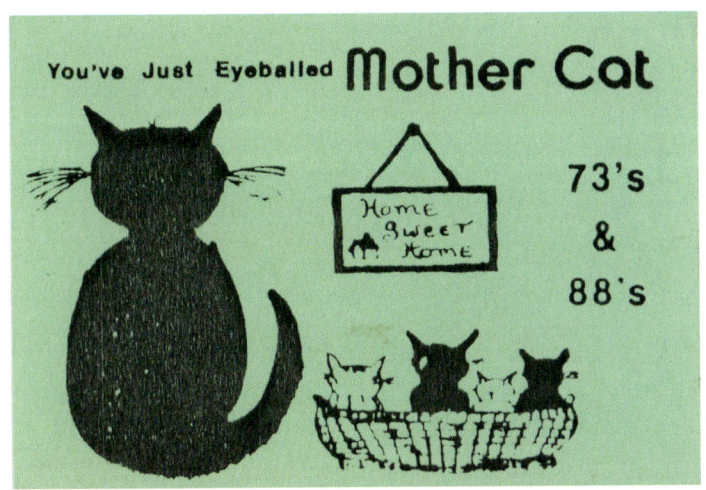

You've Just Eyeballed **Mother Cat**

Home Sweet Home

73's & 88's

EYEBALL! EYEBALL!
YOU HAVE JUST EYEBALLED

HOUND DOG
CASTLETOWN BREAKER

10-10 Till We Break Again

EYEBALL   EYEBALL

You Have Just Eyeballed

ANIMAL

Rock on we're gone bye

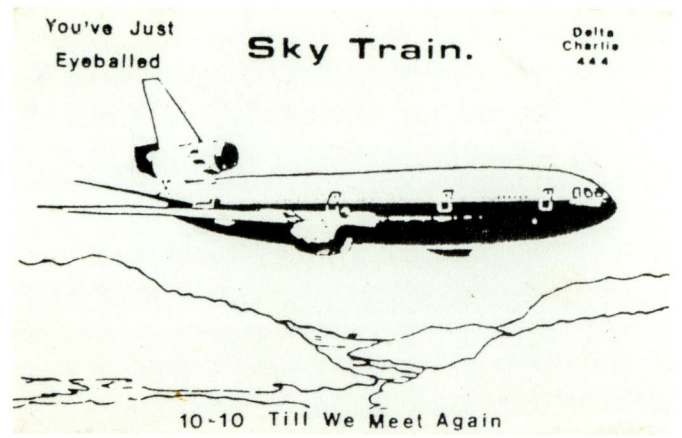

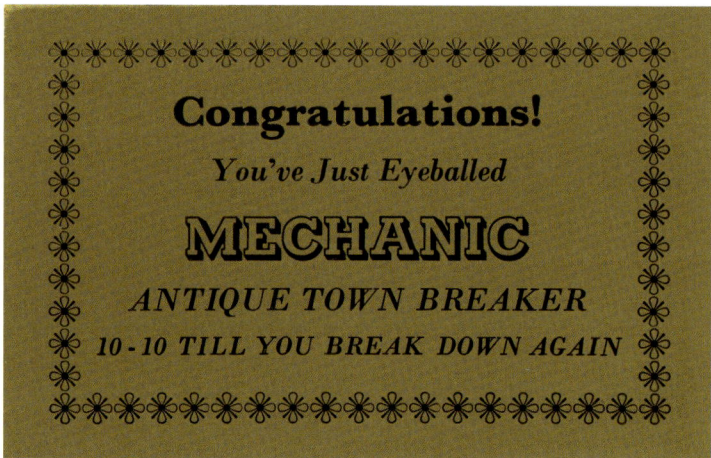

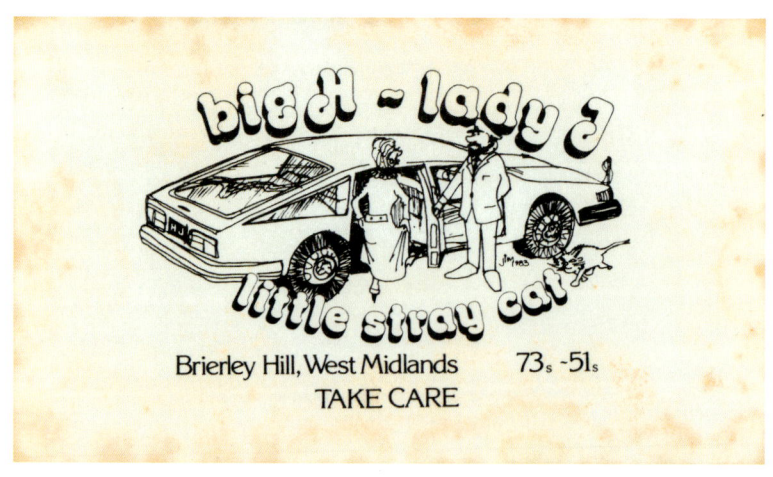

Carabina One Alex

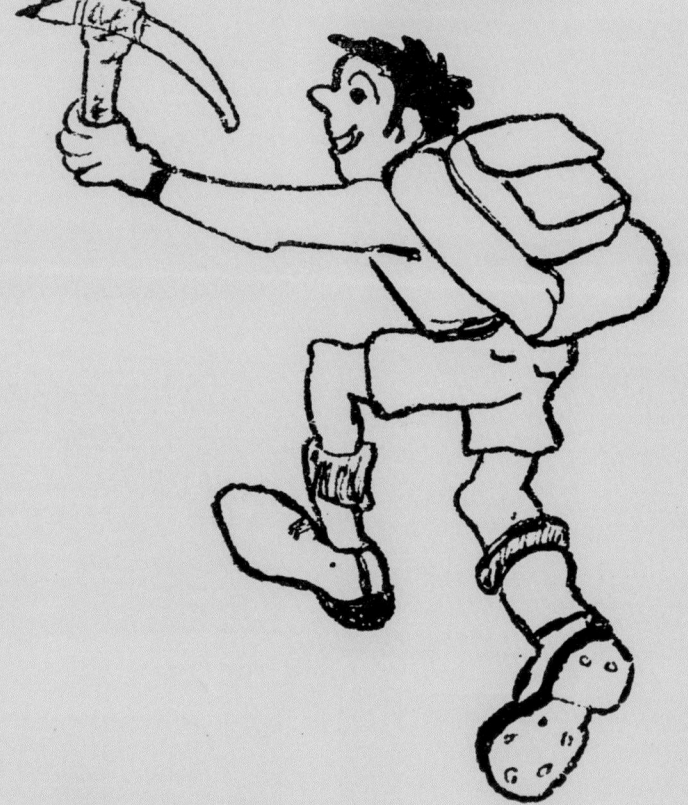

Hastings 20

EYEBALL   EYEBALL
You Have Just Eyeballed

CHICKEN HAWK

WORLINGWORTH BREAKER

10—10 Till we meet again

Eyeball   Eyeball
You've just eyeballed
'CINZANO'
(Border   Breakers)
10-10 Till we break again

EYEBALL! EYEBALL!

You Have Just Eyeballed

CAVALIER

Lawford Breaker

10-10 Till We Break Again

EYEBALL! EYEBALL!

YOU HAVE JUST EYEBALLED

NUTTY SLACK

DEBENHAM BREAKER

10 - 10 TILL WE DO IT AGAIN

EYEBALL   EYEBALL
You Have Just Eyeballed

DRAGON FLY

Castle Town Breaker
10 - 10 Till We Meet Again

EYEBALL! EYEBALL!

You Have Just Eyeballed

DAMAGE CASE

Otley Breaker 1-4
10-10 Till We Clash Again

EYEBALL   EYEBALL
You Have Just Eyeballed

* * * * * * * * * * *
WOLFIE
* * * * * * * * * * *

1 - 40 Don't be naughty

EYEBALL! EYEBALL!

You Have Just Eyeballed

FLYING CHIPPIE

10-10 Till We Break Again

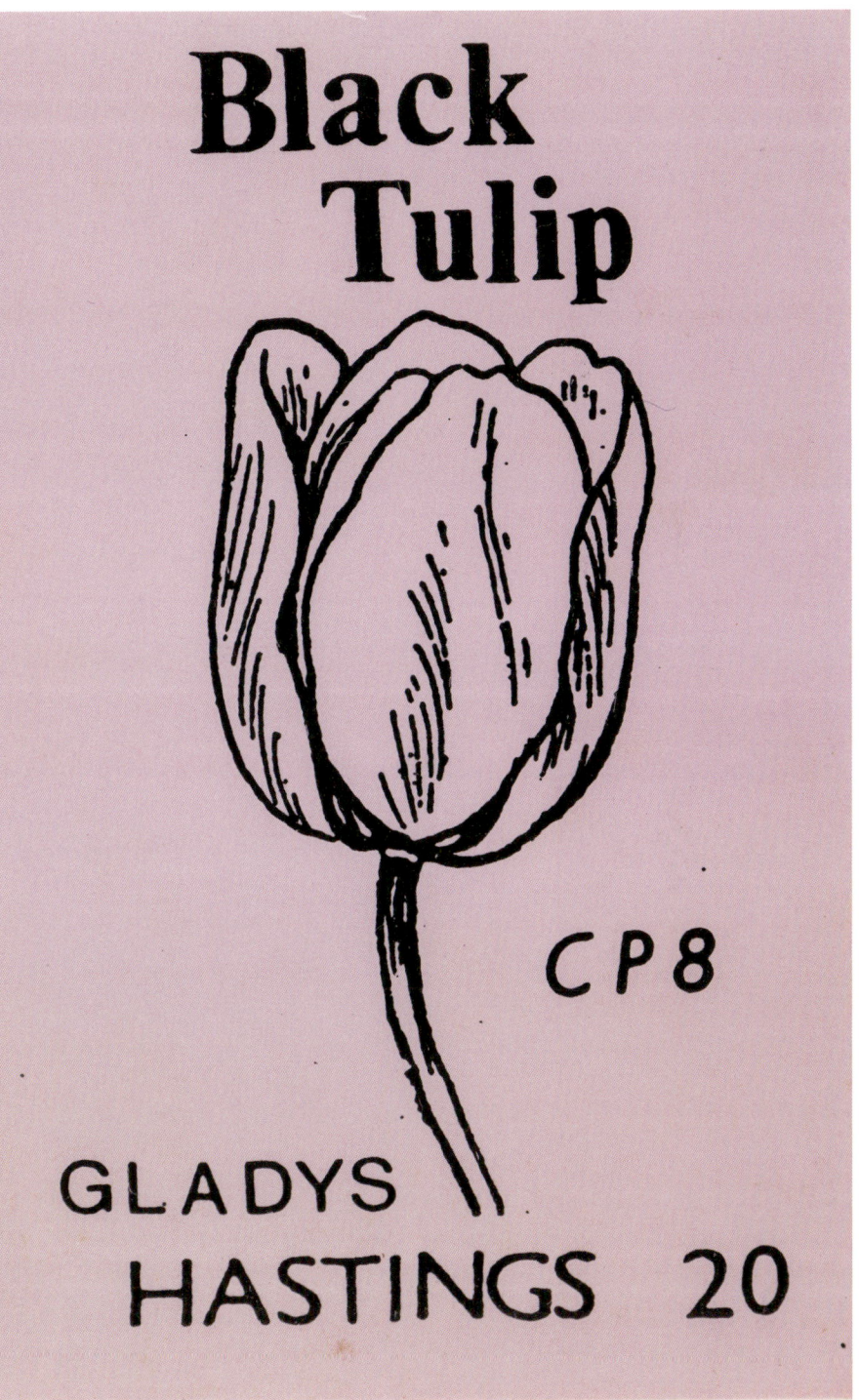

Black Tulip

CP8

GLADYS HASTINGS 20

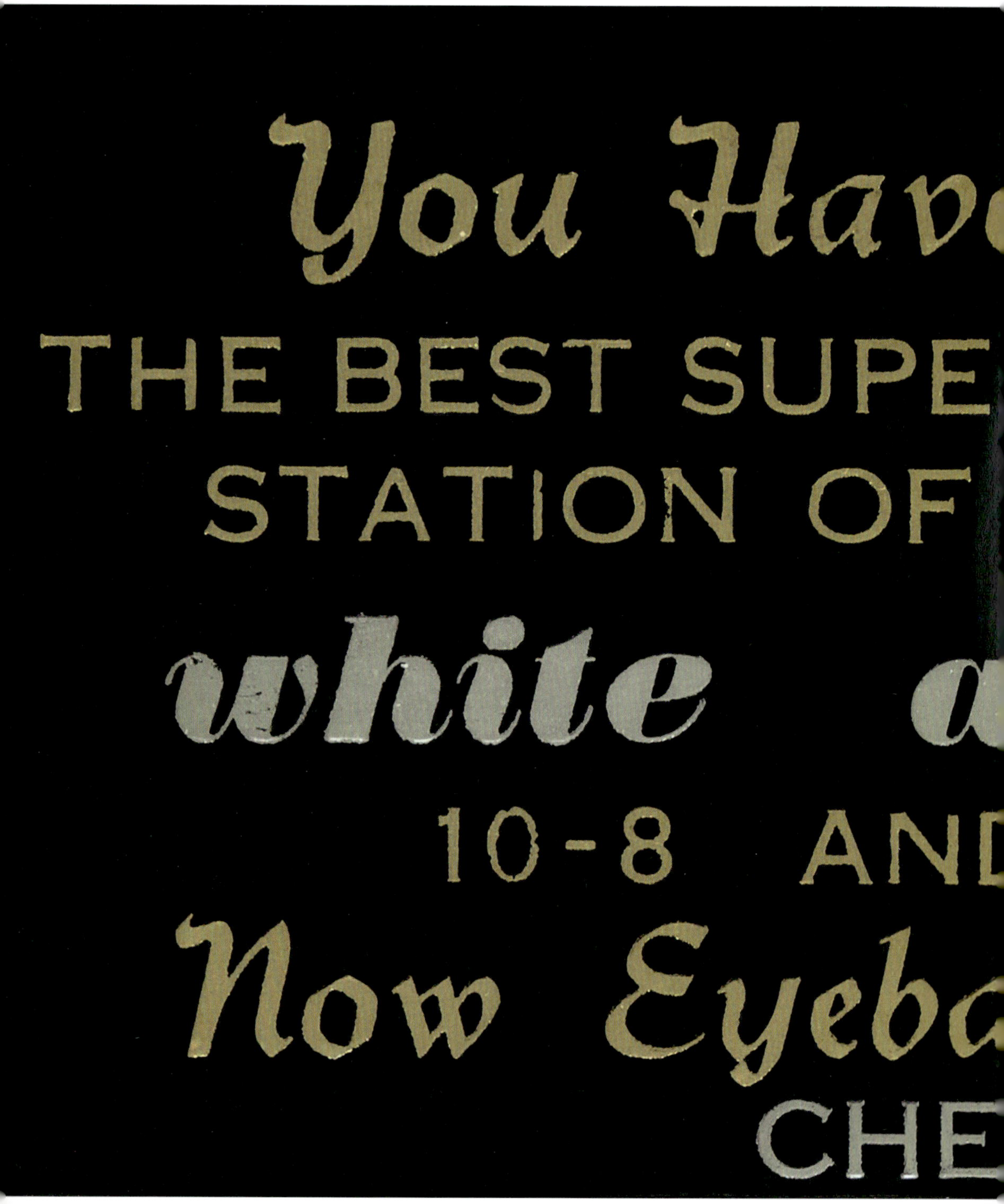

You Hav[e]

THE BEST SUPE[R]

STATION OF

white *a[nd]*

10-8   AND

Now Eyeba[ll]

CHE[...]

Eyeballed

R MODULAT!NG

THE NATION

lbatross

CLEAR

ll the Rest

ERO

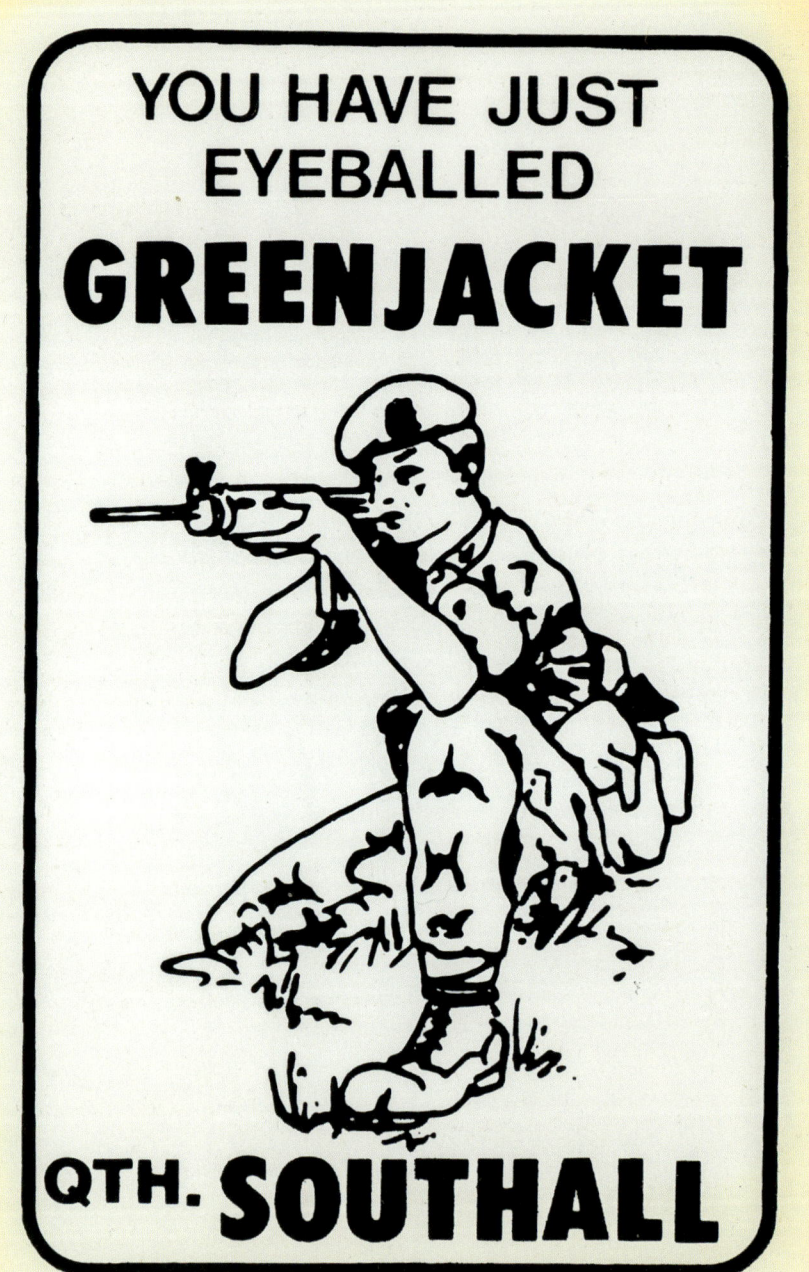

YOU HAVE JUST EYEBALLED

GREENJACKET

QTH. SOUTHALL

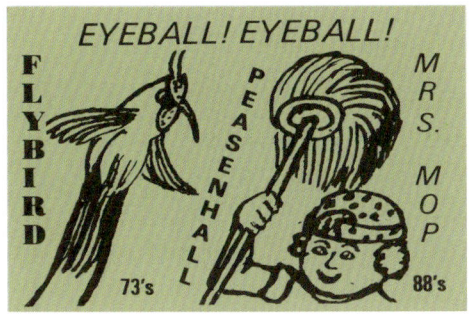

EYEBALL! EYEBALL!
FLYBIRD
PEASENHALL
MRS. MOP
73's
88's

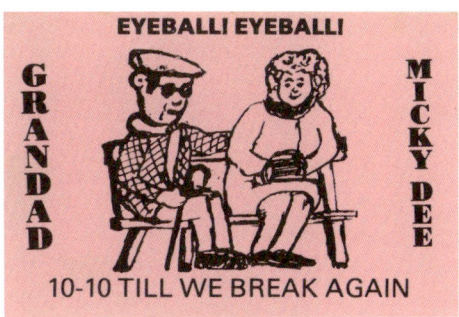

EYEBALL! EYEBALL!
GRANDAD
MICKY DEE
10-10 TILL WE BREAK AGAIN

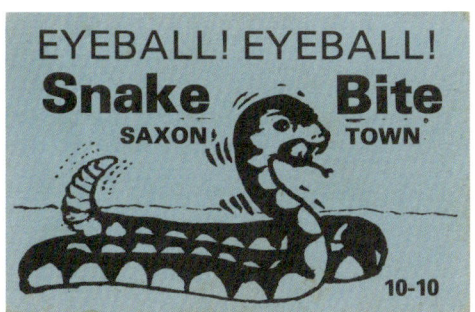

EYEBALL! EYEBALL!
Snake Bite
SAXON
TOWN
10-10

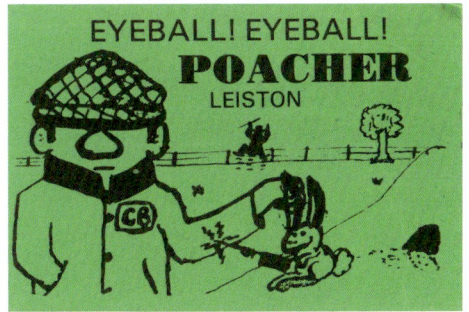

EYEBALL! EYEBALL!
POACHER
LEISTON
CB

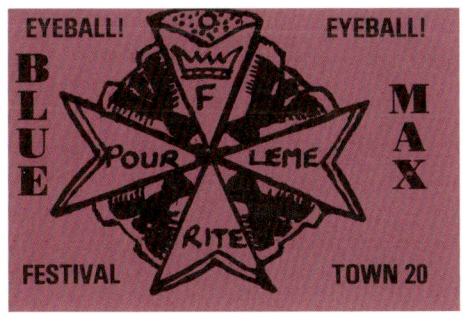

EYEBALL!
EYEBALL!
BLUE
MAX
F
POUR
LEME
RITE
FESTIVAL
TOWN 20

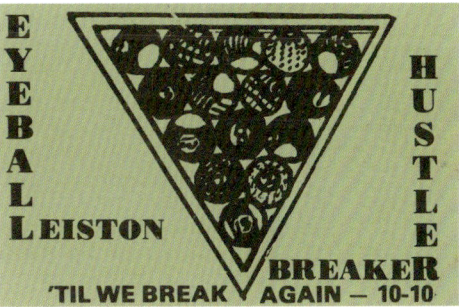

EYEBALL
LEISTON
HUSTLE
BREAKER
'TIL WE BREAK
AGAIN — 10-10

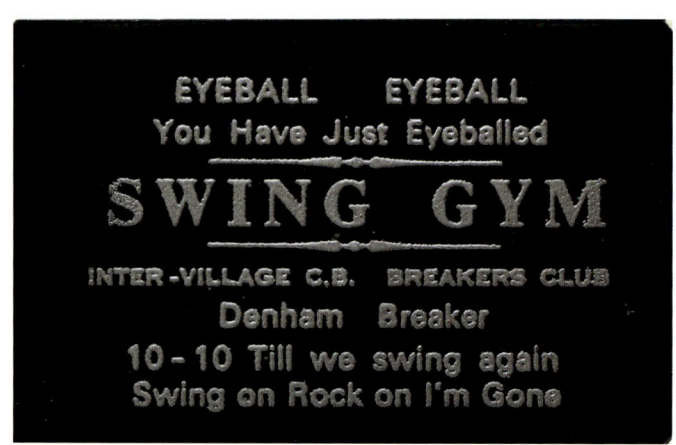

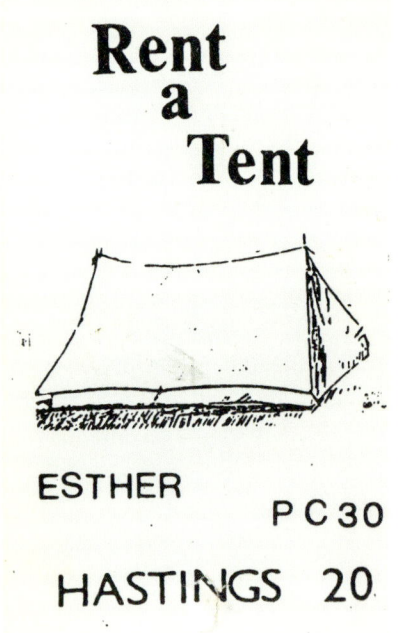

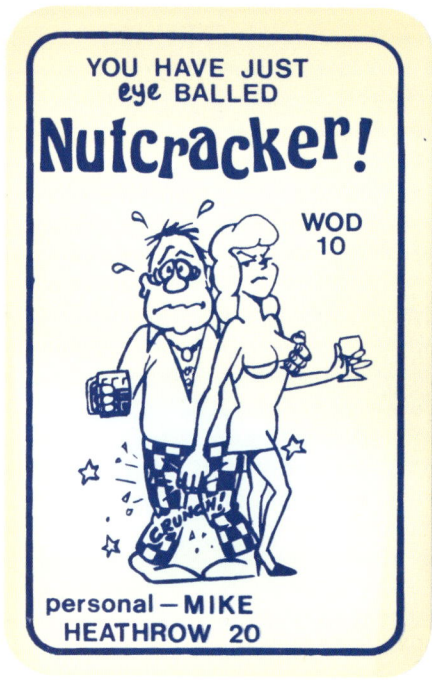

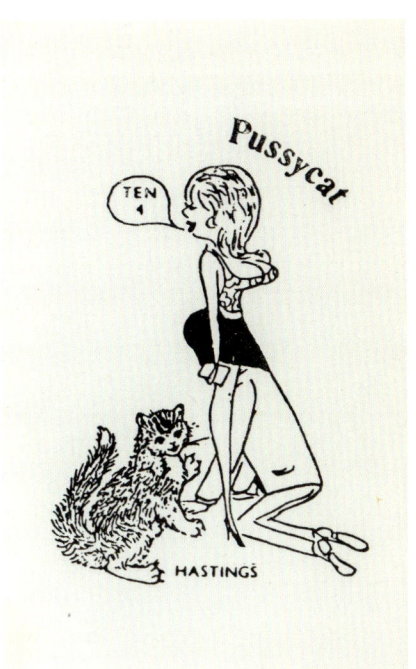

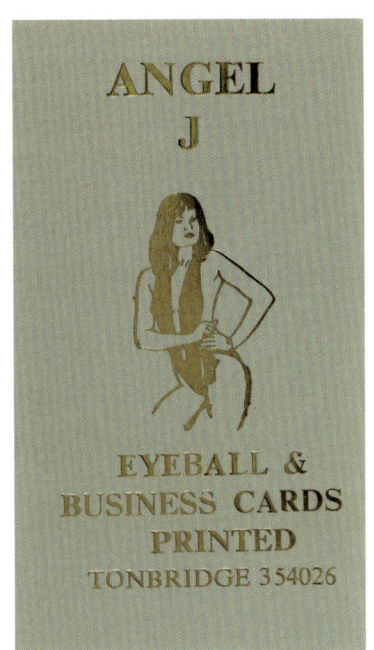

YOU
JUST EYEBALLED

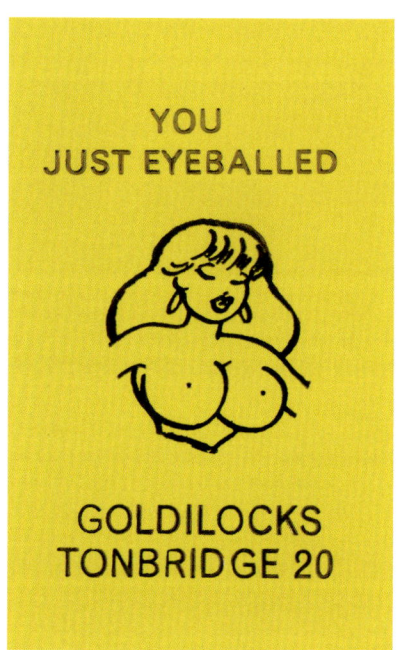

GOLDILOCKS
TONBRIDGE 20

CONGRATULATIONS
You've Just Eyeballed
SNOW WHITE
Timber Town Breaker
10-10 Till We Break Again

GRANDAD ROY

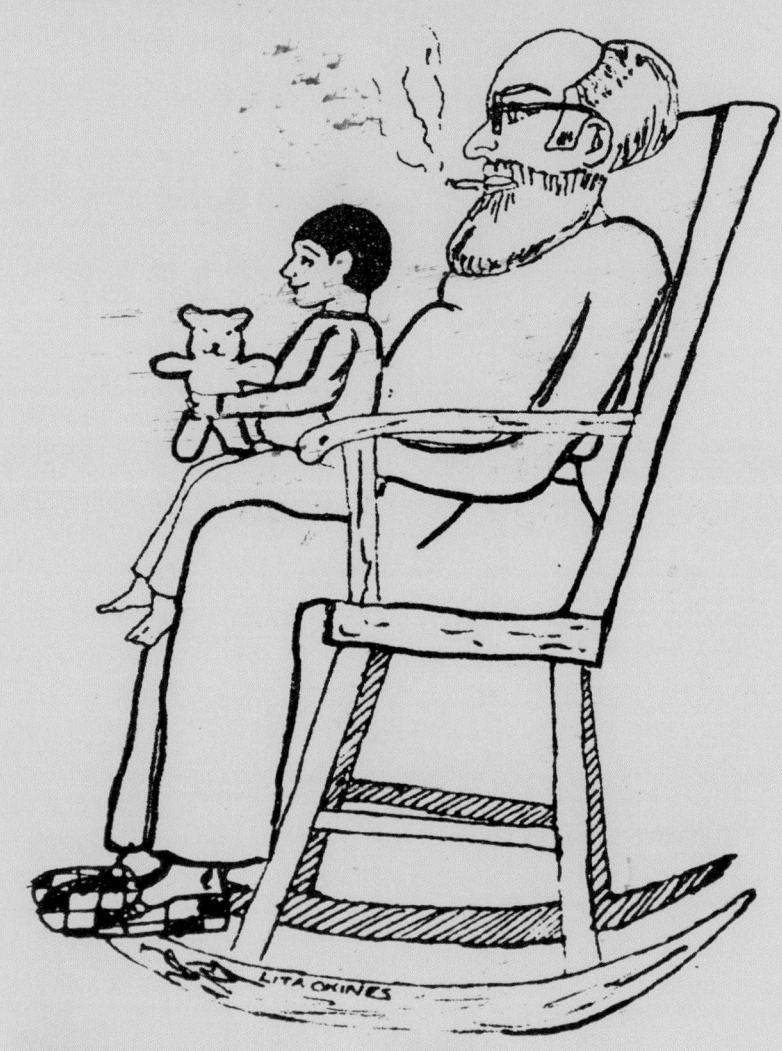

LITA OKINES

Bexhill 20

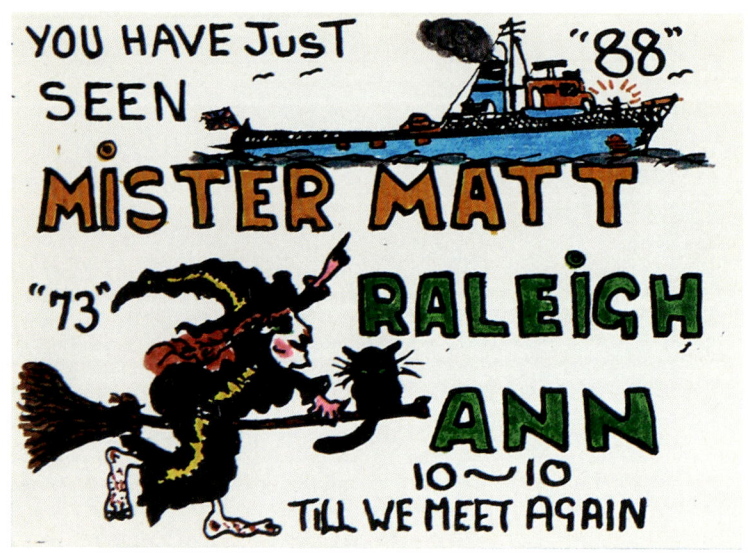

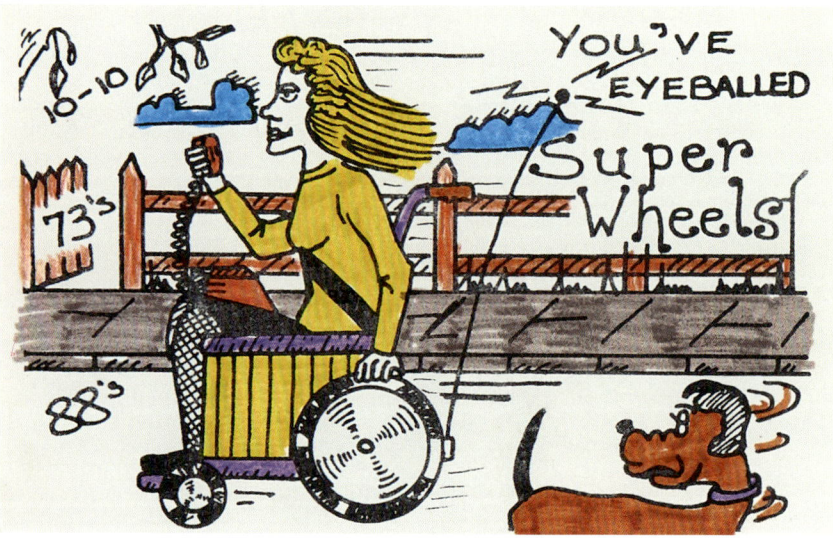

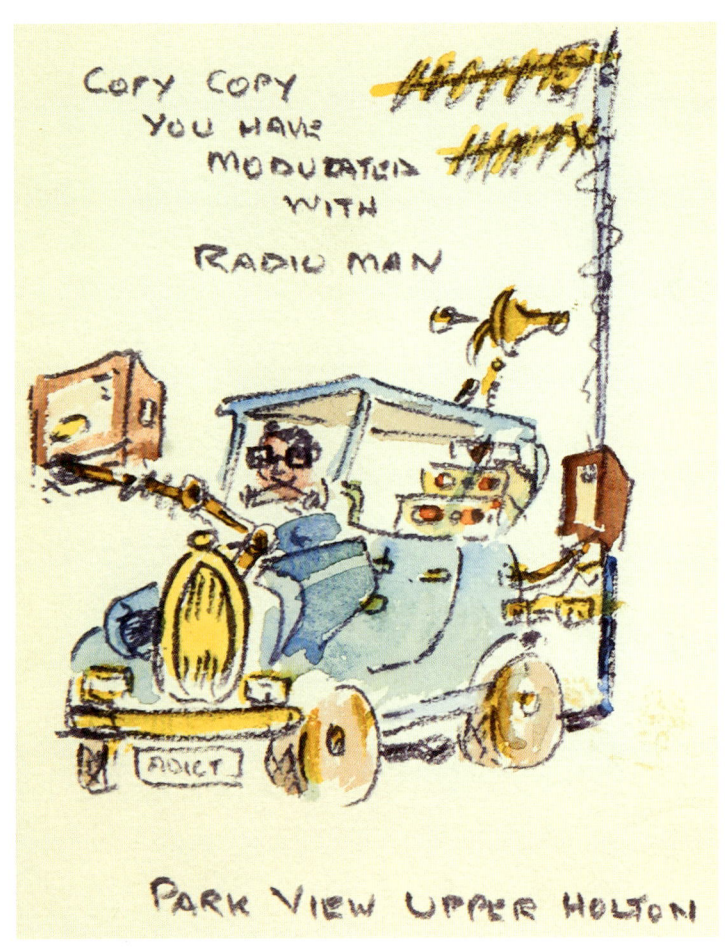

You have just been
Eyeballed by

RED DUCK

SUNDRIDGE   PARK

10–10 till we meet again

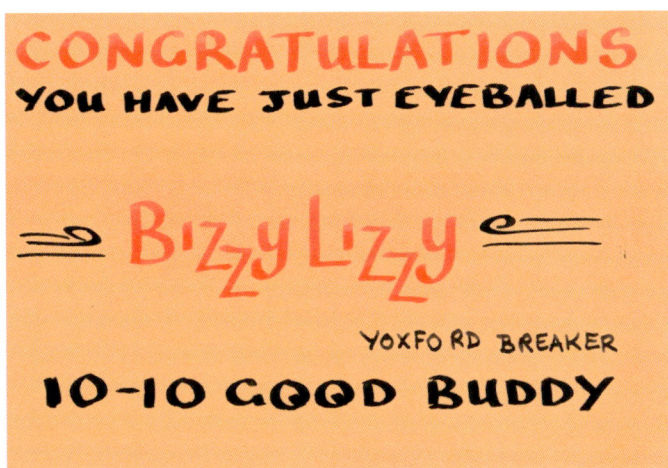

CONGRATULATIONS
YOU HAVE JUST EYEBALLED

Bizzy Lizzy

YOXFORD BREAKER

10-10 GOOD BUDDY

EYEBALL! EYEBALL!
You Have Just Eyeballed

Straight Eight & Bissy Lizzy

Super Modulator
Express Breaker

You Have Been Struck By

LIGHTNIN'

From The Ulcombe Twenty
Tel: 0622-842632

EYEBALL! EYEBALL!

You Have Just Eyeballed

SKY GOD

10-10 Till We Break Again

YOU HAVE JUST EYEBALLED

Miami   Dolphin

Ivor

Unit 43   P.O.   Box 5
Ammanford,   Dyfed. S A 18 3 B N
ALL THE BEST TO YOU & YOURS

EYEBALL! EYEBALL!
You've Just Eyeballed
SHEPPIE
10-10 Till We Break Again

EYEBALL   EYEBALL
You Have Just Eyeballed
Dream Topping
10-10 Till We Dream Again

You Have

Eyeballed

MURRY MINT

ROBERT

Hastings 20

EYEBALL! EYEBALL!

OSCAR 9
TUNGSTON TIP
YOXFORD BREAKER

10-10 TILL THE NEXT TIME

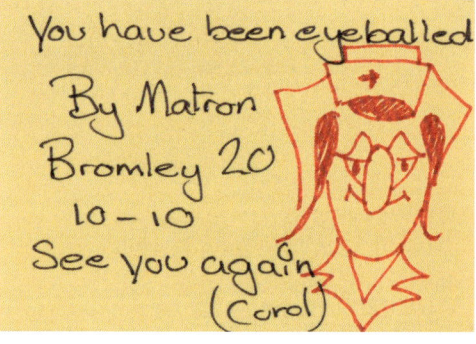

You have been eyeballed

By Matron
Bromley 20
10 - 10
See you again
(Carol)

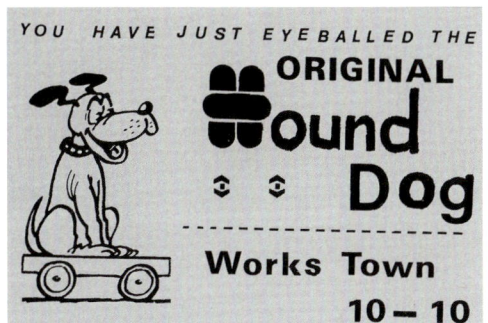

YOU HAVE JUST EYEBALLED THE
ORIGINAL
Hound
Dog
Works Town
10 – 10

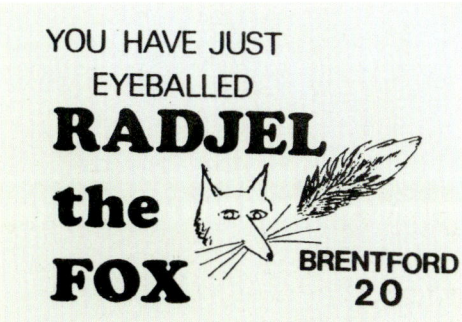

YOU HAVE JUST
EYEBALLED
RADJEL
the
FOX
BRENTFORD
20

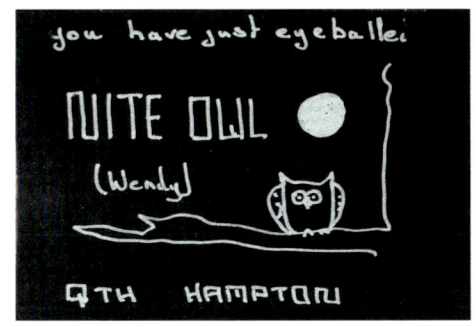

you have just eyeballed
NITE OWL
(Wendy)
9TH HAMPTON

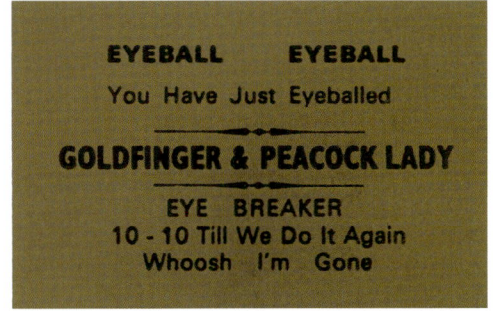

EYEBALL   EYEBALL
You Have Just Eyeballed

GOLDFINGER & PEACOCK LADY

EYE BREAKER
10 - 10 Till We Do It Again
Whoosh I'm Gone

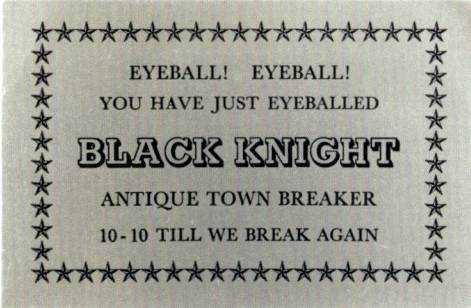

EYEBALL! EYEBALL!
YOU HAVE JUST EYEBALLED

BLACK KNIGHT

ANTIQUE TOWN BREAKER

10 - 10 TILL WE BREAK AGAIN

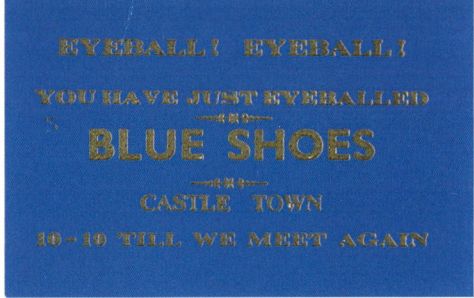

EYEBALL! EYEBALL!
YOU HAVE JUST EYEBALLED
BLUE SHOES
CASTLE TOWN
10 - 10 TILL WE MEET AGAIN

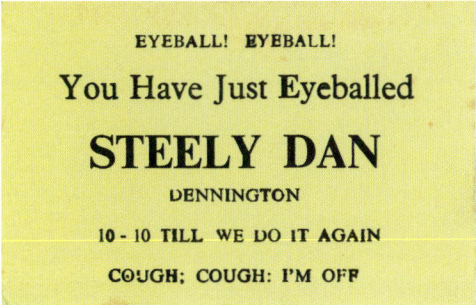

EYEBALL! EYEBALL!
You Have Just Eyeballed

STEELY DAN

DENNINGTON

10 - 10 TILL WE DO IT AGAIN

COUGH; COUGH: I'M OFF

EYEBALL   EYEBALL
You Have Just Eyeballed

* * * * * * * * * * *
SKID LID
* * * * * * * * * * *

10-10 TILL WE FLY AGAIN

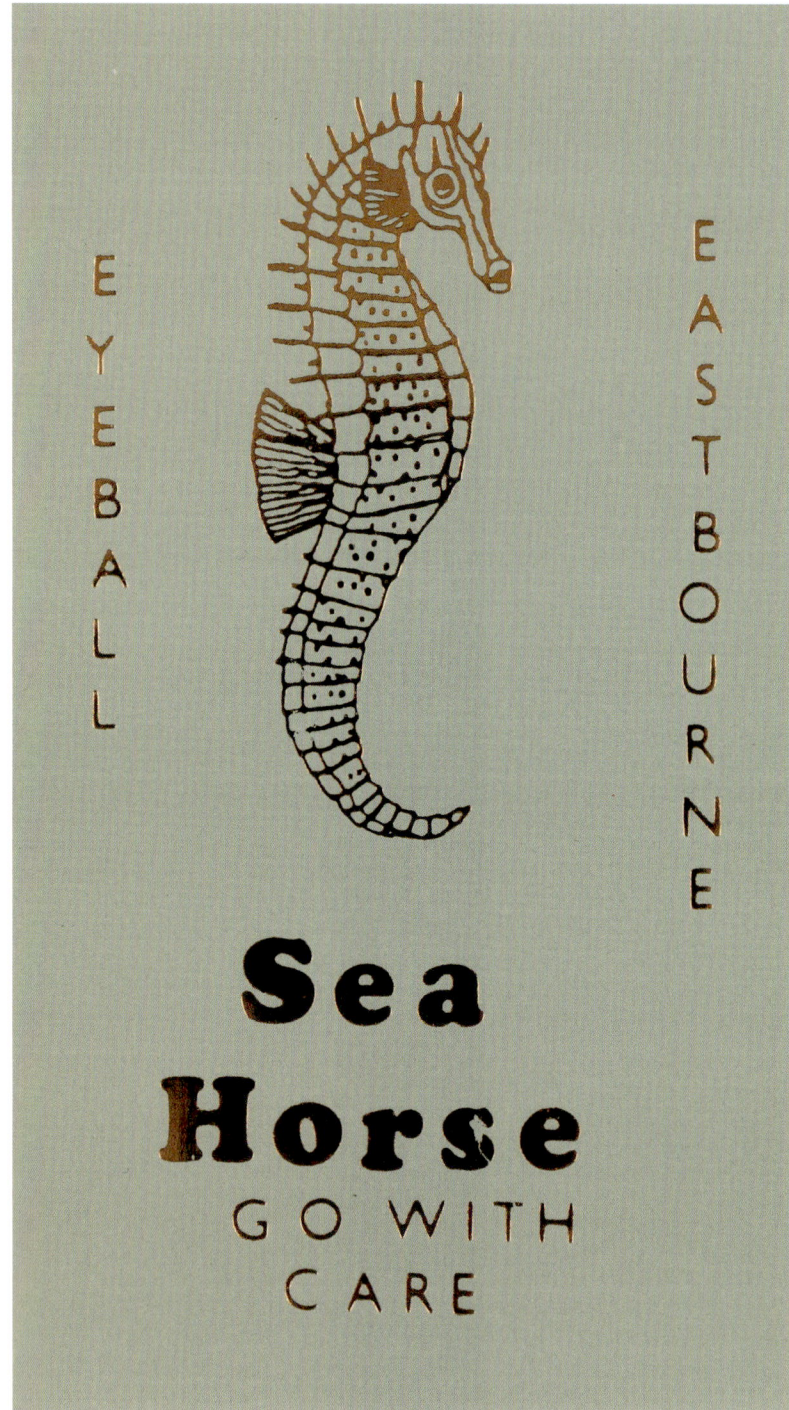

EYEBALL

EASTBOURNE

Sea
Horse
GO WITH
CARE

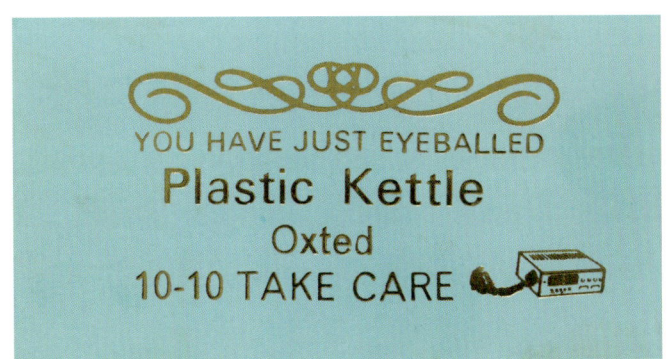

YOU HAVE JUST EYEBALLED
Plastic Kettle
Oxted
10-10 TAKE CARE

EYEBALL ! EYEBALL !

You Have Just Eyeballed

Big Lady & Big G

Great Glemham Breaker

10 - 10 Till We Do It Again

Take Care Because We Both Care

CONGRATULATIONS!
YOU HAVE JUST EYEBALLED

SPECIAL K

Yoxford Breaker

10-10 TILL WE BREAK AGAIN

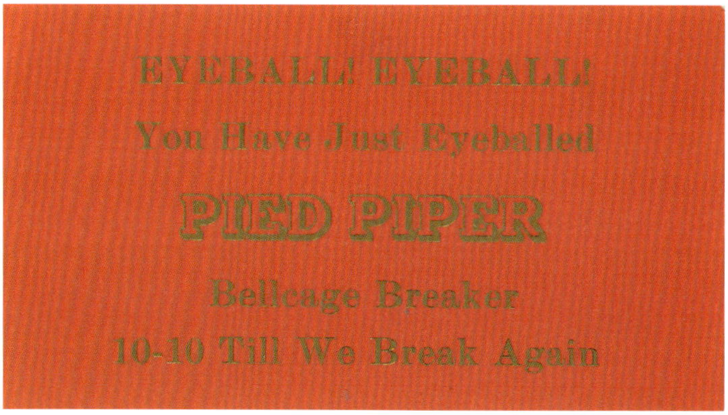

EYEBALL! EYEBALL!

You Have Just Eyeballed

PIED PIPER

Bellcage Breaker

10-10 Till We Break Again

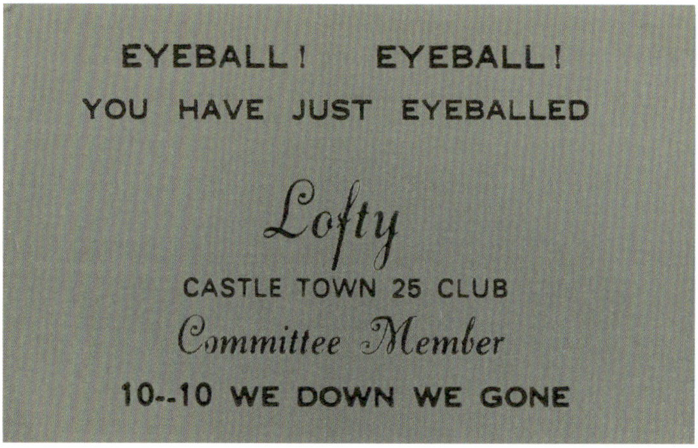

EYEBALL!   EYEBALL!

YOU HAVE JUST EYEBALLED

Lofty

CASTLE TOWN 25 CLUB

Committee Member

10--10 WE DOWN WE GONE

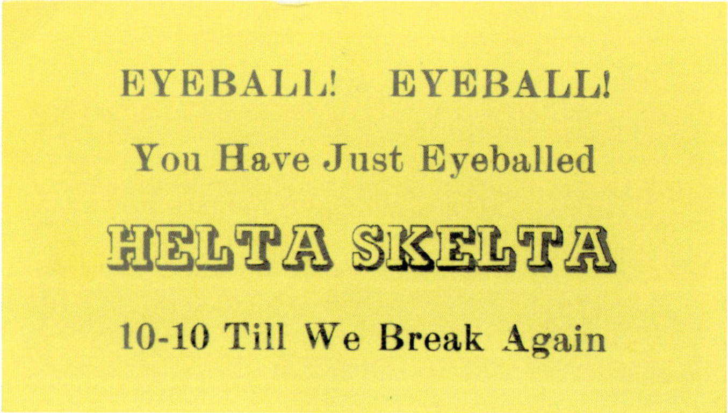

EYEBALL!   EYEBALL!

You Have Just Eyeballed

HELTA SKELTA

10-10 Till We Break Again

Foxy Fred

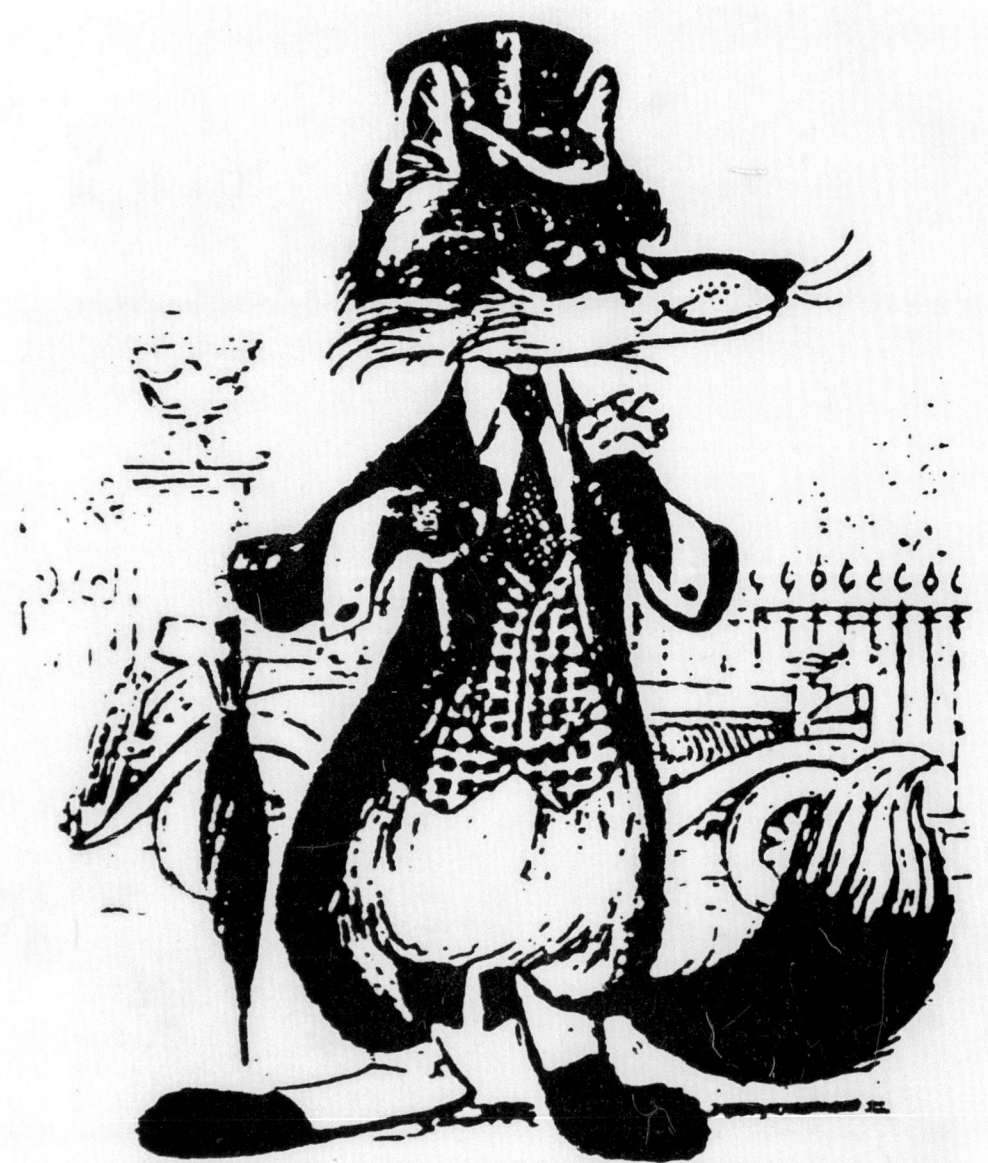

HASTINGS

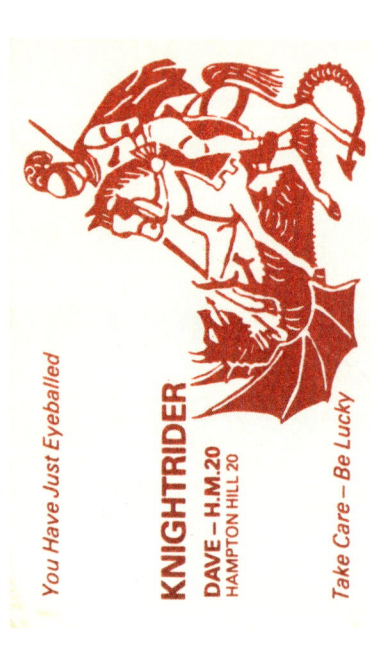

You Have Just Eyeballed

KNIGHTRIDER
DAVE – H.M.20
HAMPTON HILL 20

Take Care – Be Lucky

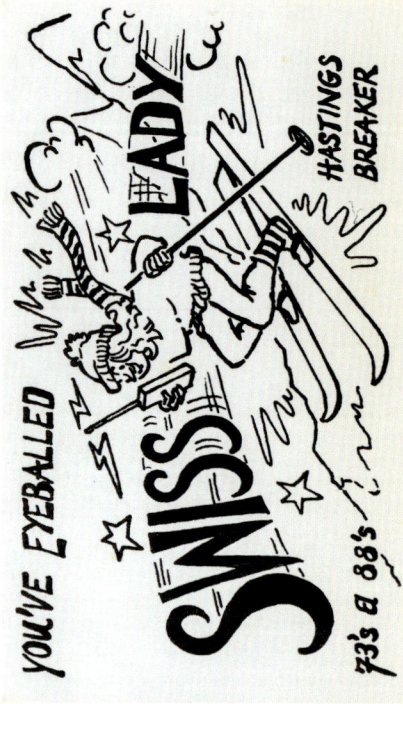

YOU'VE EYEBALLED

SWISS LADY

HASTINGS BREAKER

73's & 88's

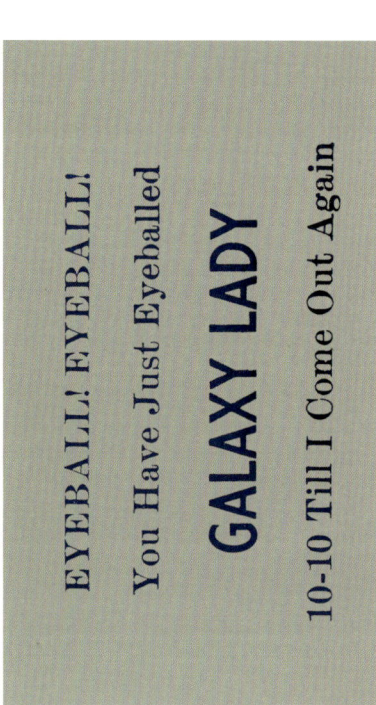

EYEBALL! EYEBALL!
You Have Just Eyeballed
GALAXY LADY
10-10 Till I Come Out Again

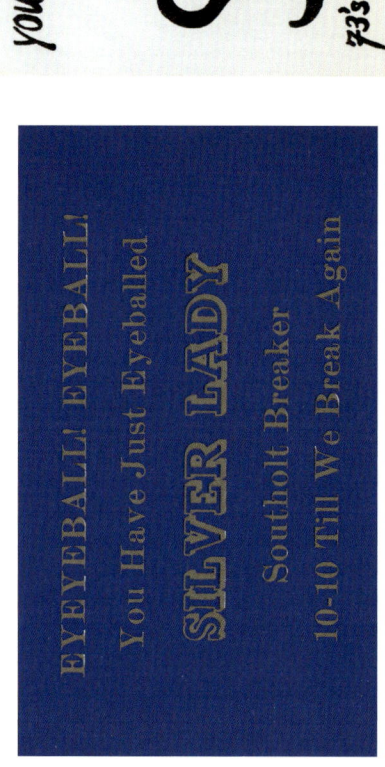

EYEYEBALL! EYEBALL!
You Have Just Eyeballed
SILVER LADY
Southolt Breaker
10-10 Till We Break Again

You Have Just Eyeballed

Hammer Handle
&
Dusty Bin
Hoxne Breakers
We're down we're gone

you have Just Eyeballed

SHELTiE
V.R. 290
New Elgin 20
Down
and
Gone
10 –10

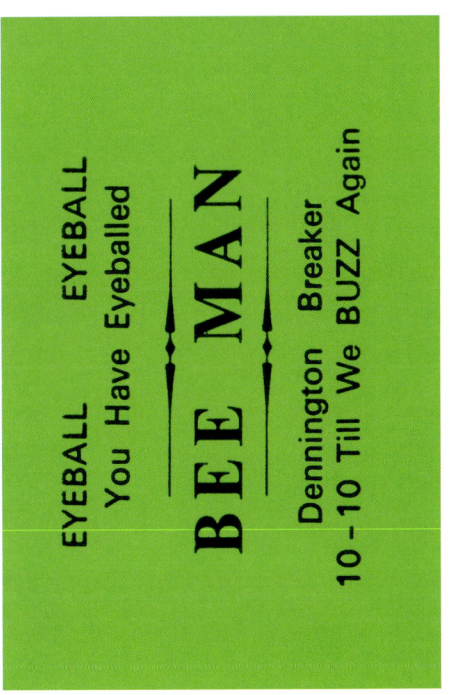

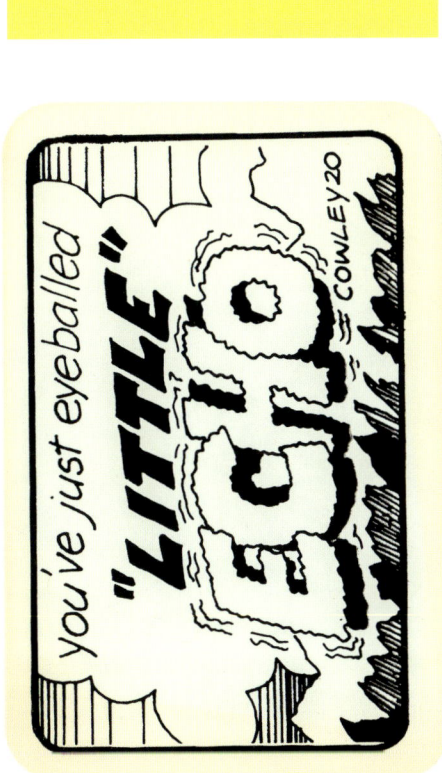

you've just eyeballed "LITTLE" ECHO
COWLEY20

EYEBALL   EYEBALL
You Have Eyeballed

BEE MAN
Dennington   Breaker
10 – 10 Till We BUZZ Again

EYEBALL
BOUN
10.10 Till we do it

EYEBALL! EYEBALL!
You Have Just Eyeballed
RED BARON
Workstown Breaker
10-10 Till We Break Again

EYEBALL! EYEBALL!
YOU HAVE JUST EYEBALLED
MINOR MAN
Cransford Breaker
10-10 Till We Shrink Again

EYEBALL! EYEBALL!
Your In The Company Of
LITTLE WITCH
Lawford 20
10-10 Till We Break Again

EYEBALL! EYEBALL!
YOU HAVE JUST EYEBALLED
SILVER LINING & DAISY
FOX'S DEN BREAKERS
1-40 DON'T BE NAUGHTY
10-10 TILL WE MEET AGAIN

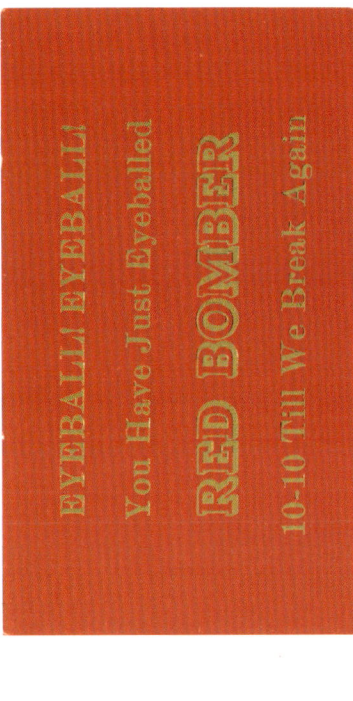

EYEBALL! EYEBALL!

You Have Just Eyeballed

RED BOMBER

10-10 Till We Break Again

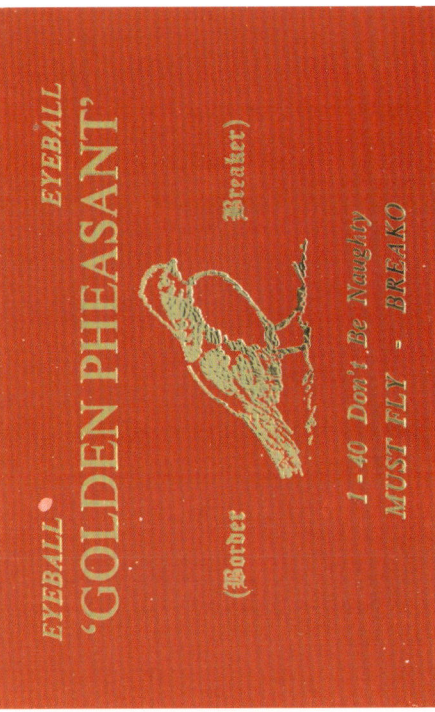

EYEBALL

'GOLDEN PHEASANT'

Breaker

EYEBALL

(Border)

1-40 Don't Be Naughty

MUST FLY - BREAKO

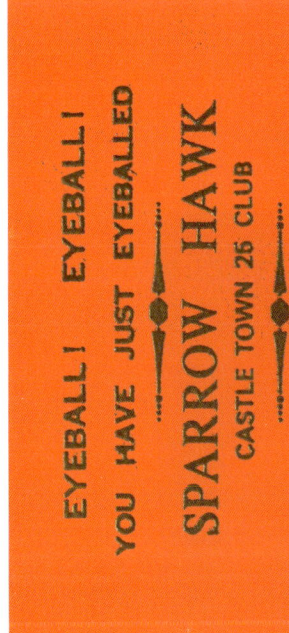

EYEBALL! EYEBALL!

YOU HAVE JUST EYEBALLED

SPARROW HAWK

CASTLE TOWN 26 CLUB

10-10 Till we do it again

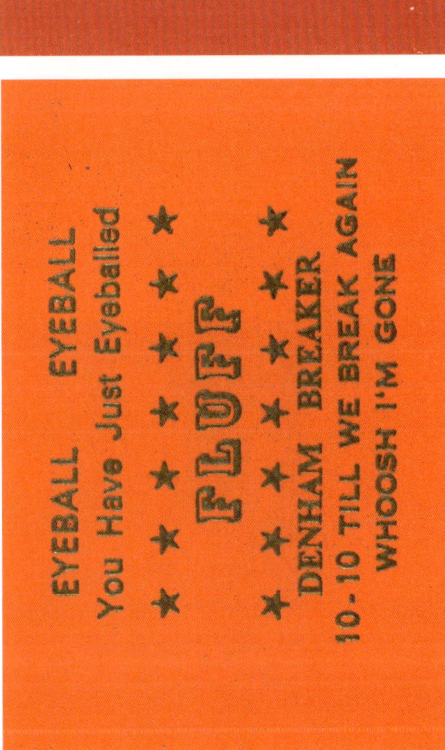

EYEBALL EYEBALL

You Have Just Eyeballed

★ ★ ★ ★ ★ ★

★ ★ FLUFF ★ ★

★ ★ DENHAM BREAKER ★ ★

10-10 TILL WE BREAK AGAIN

WHOOSH I'M GONE

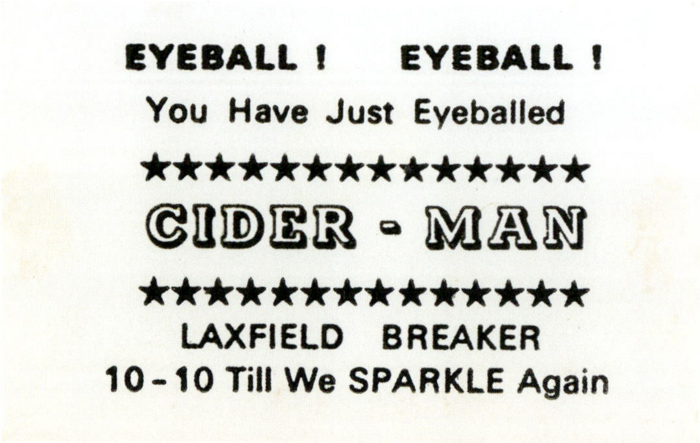

EYEBALL !   EYEBALL !

You Have Just Eyeballed

★★★★★★★★★★★★★★

CIDER - MAN

★★★★★★★★★★★★★★

LAXFIELD   BREAKER

10 - 10 Till We SPARKLE Again

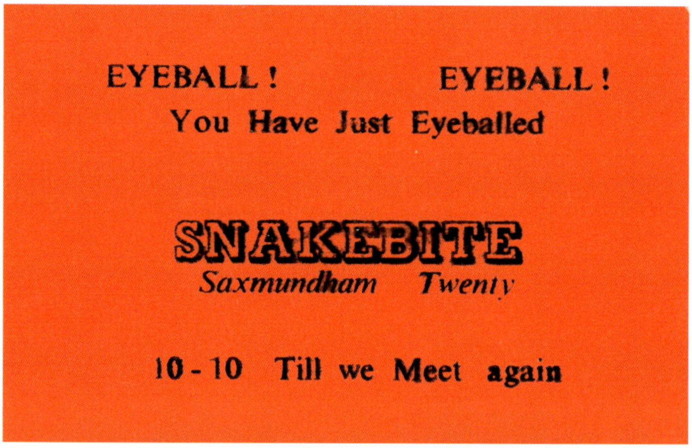

EYEBALL !   EYEBALL !

You Have Just Eyeballed

SNAKEBITE

Saxmundham   Twenty

10 - 10   Till we Meet again

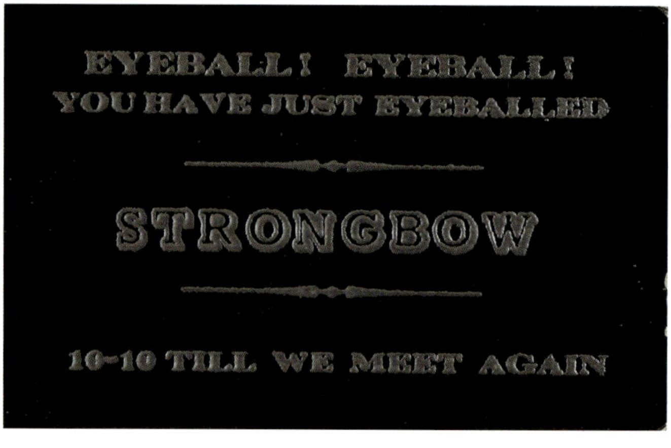

EYEBALL ! EYEBALL !
YOU HAVE JUST EYEBALLED

STRONGBOW

10-10 TILL WE MEET AGAIN

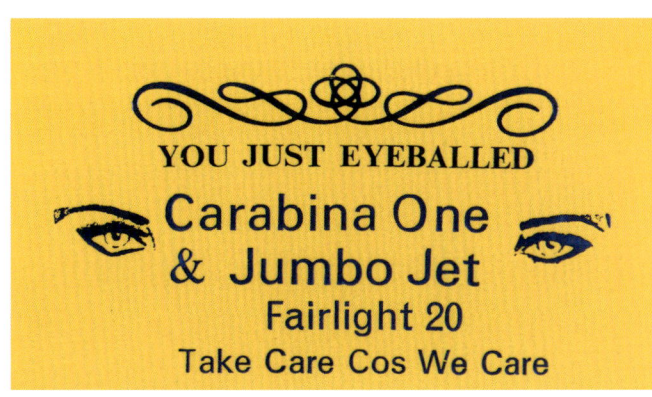

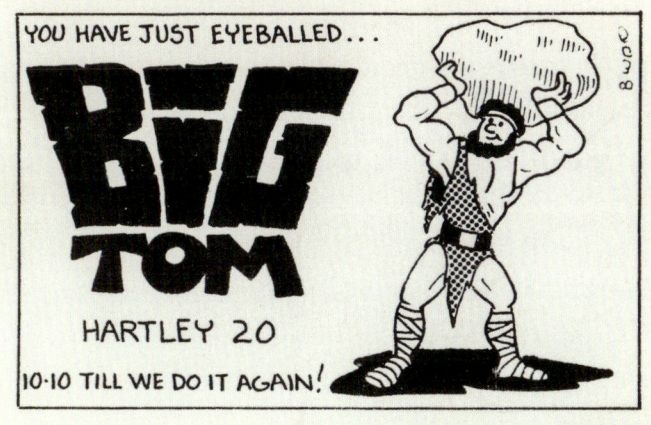

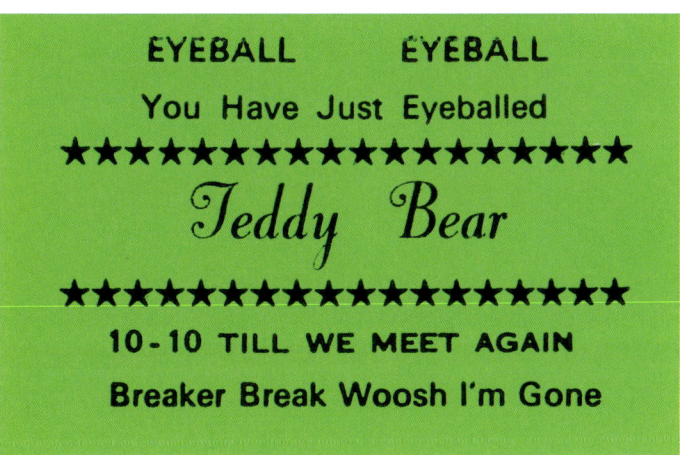

> **Eyeball**   **Eyeball**
> **YOU HAVE JUST EYEBALLED**
>
> **DART PLAYER**
>
> **Works Town Breaker**
> **10-10 TILL WE BREAK AGAIN**

Peter Hurren, aka 'Dart Player'.

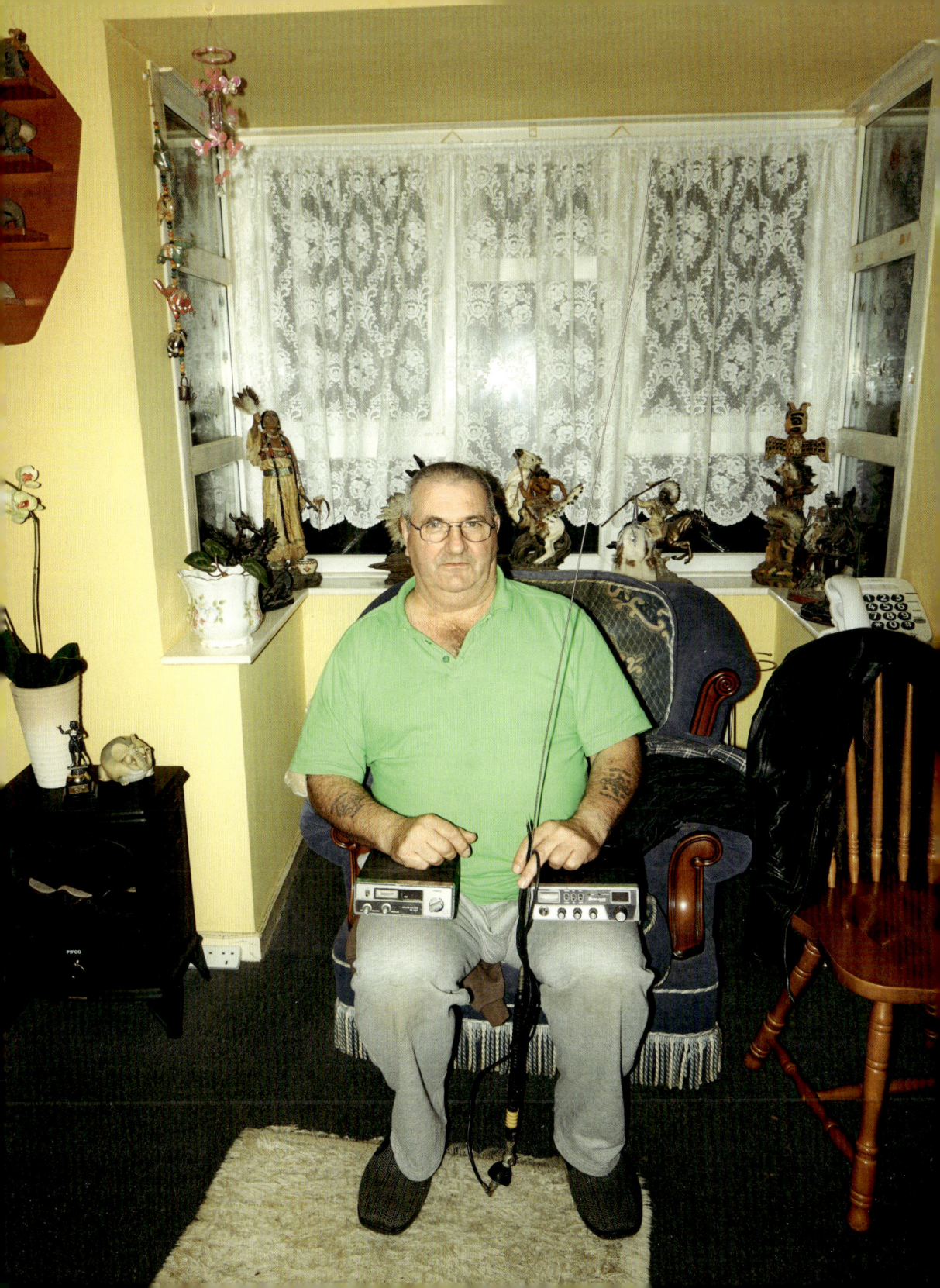

CONGRATULATIONS!

YOU HAVE JUST EYEBALLED

MUTLEY &
MISS PIGGY

Works Town Breakers

10-10 TILL WE BREAK AGAIN

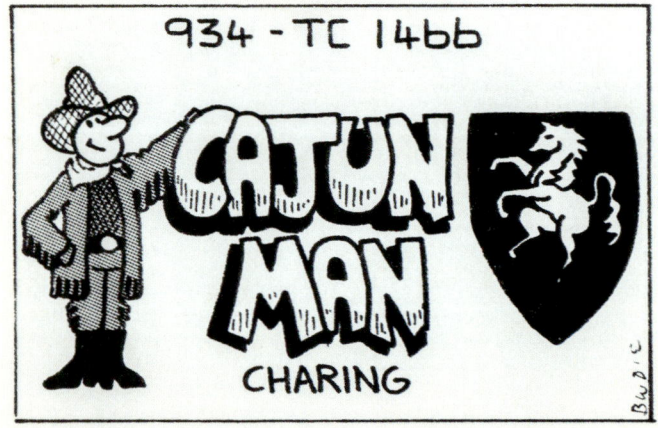

934 - TC 1466

CAJUN MAN

CHARING

CONGRATULATIONS!

YOU HAVE JUST EYEBALLED

HOT WHEELS

Dennington Breaker

10-10 TILL WE BREAK AGAIN

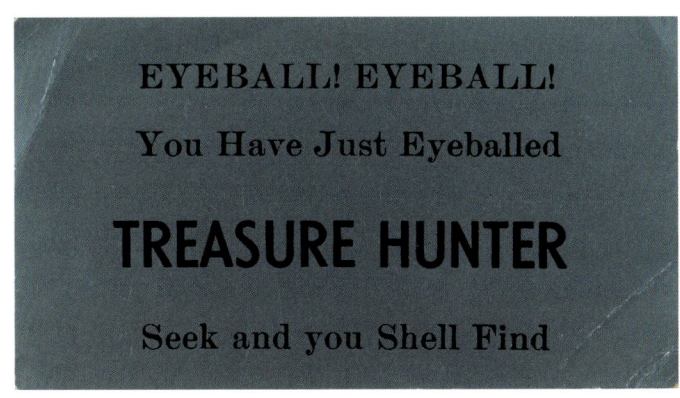

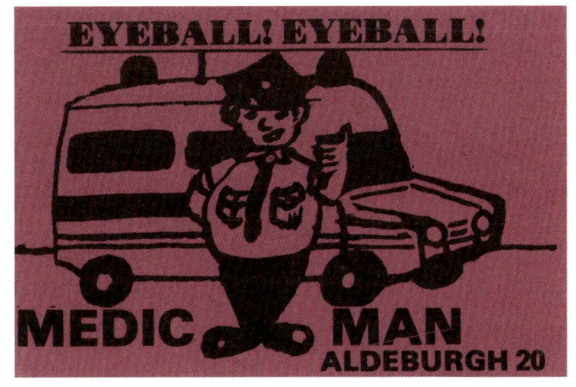

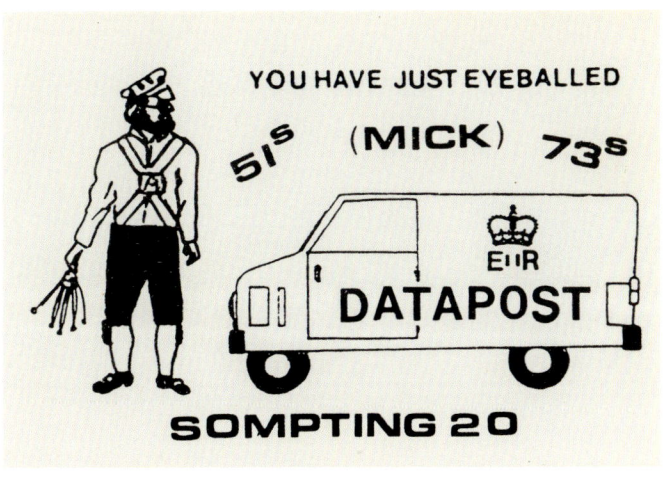

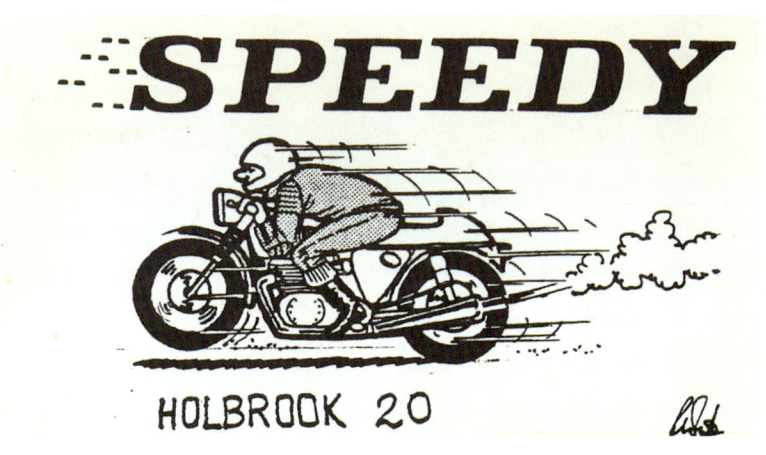

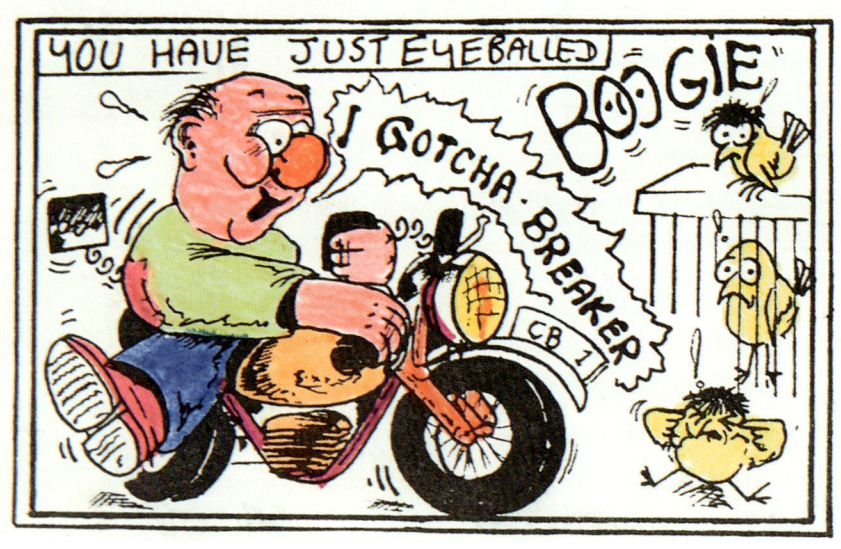

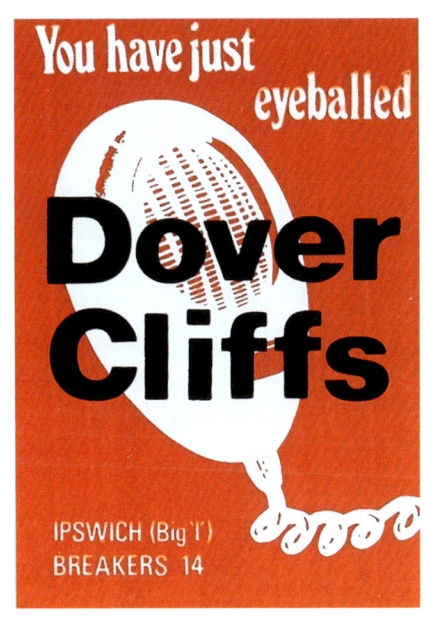

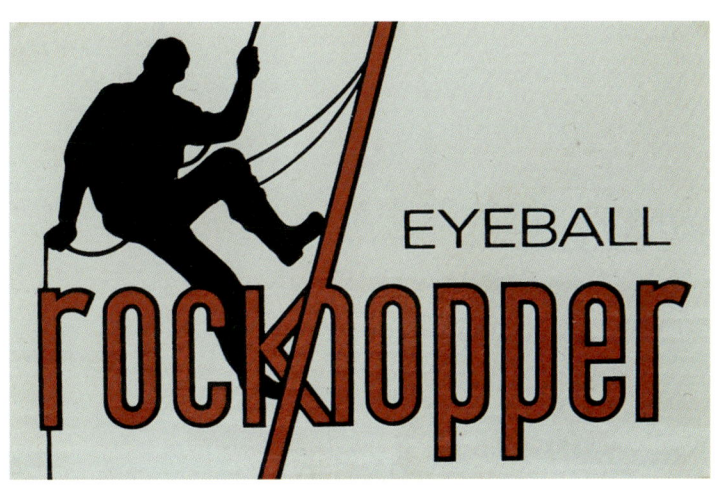

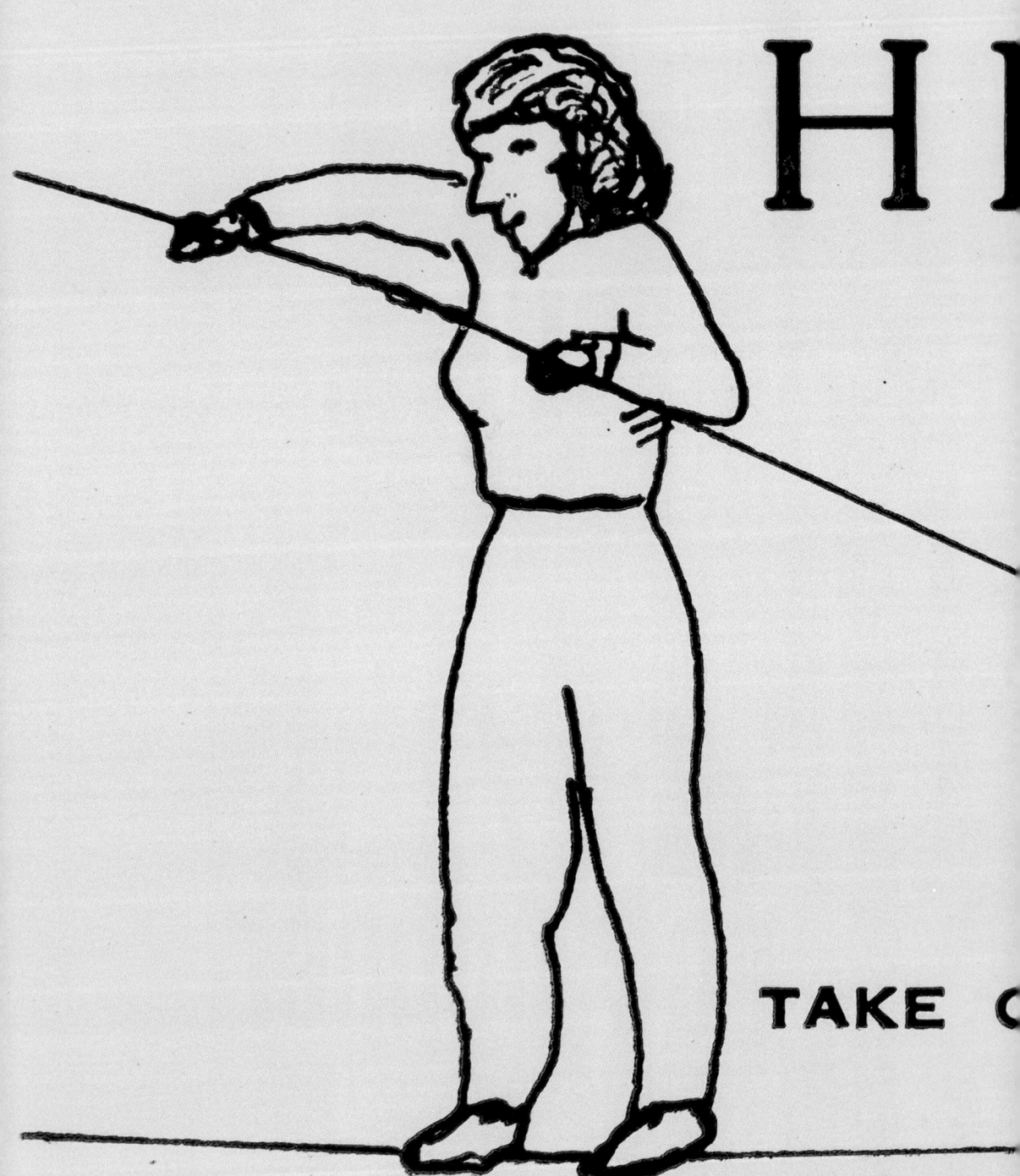

HE

TAKE C

-JEAN

BROMLEY

1st PERSONAL

JEAN

ARE BECAUSE WE CARE

EYEBALL ! EYEBALL !
You Have Just Eyeballed
★★★★★★★★★★★★★★

Mucky Duck

★★★★★★★★★★★★★★★★
CRATFIELD   BREAKER
10 - 10 Till We Break Again

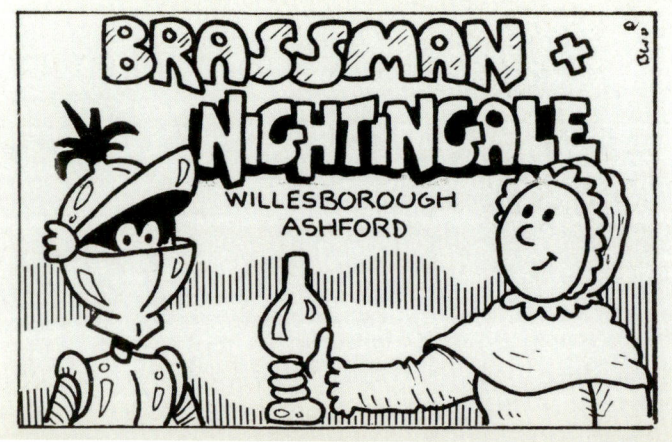

EYEBALL! EYEBALL!

YOU HAVE JUST EYEBALLED

**HONDA LADY
& SIDE CAR**
CRANSFORD BREAKER

10-10 Till We Break Again

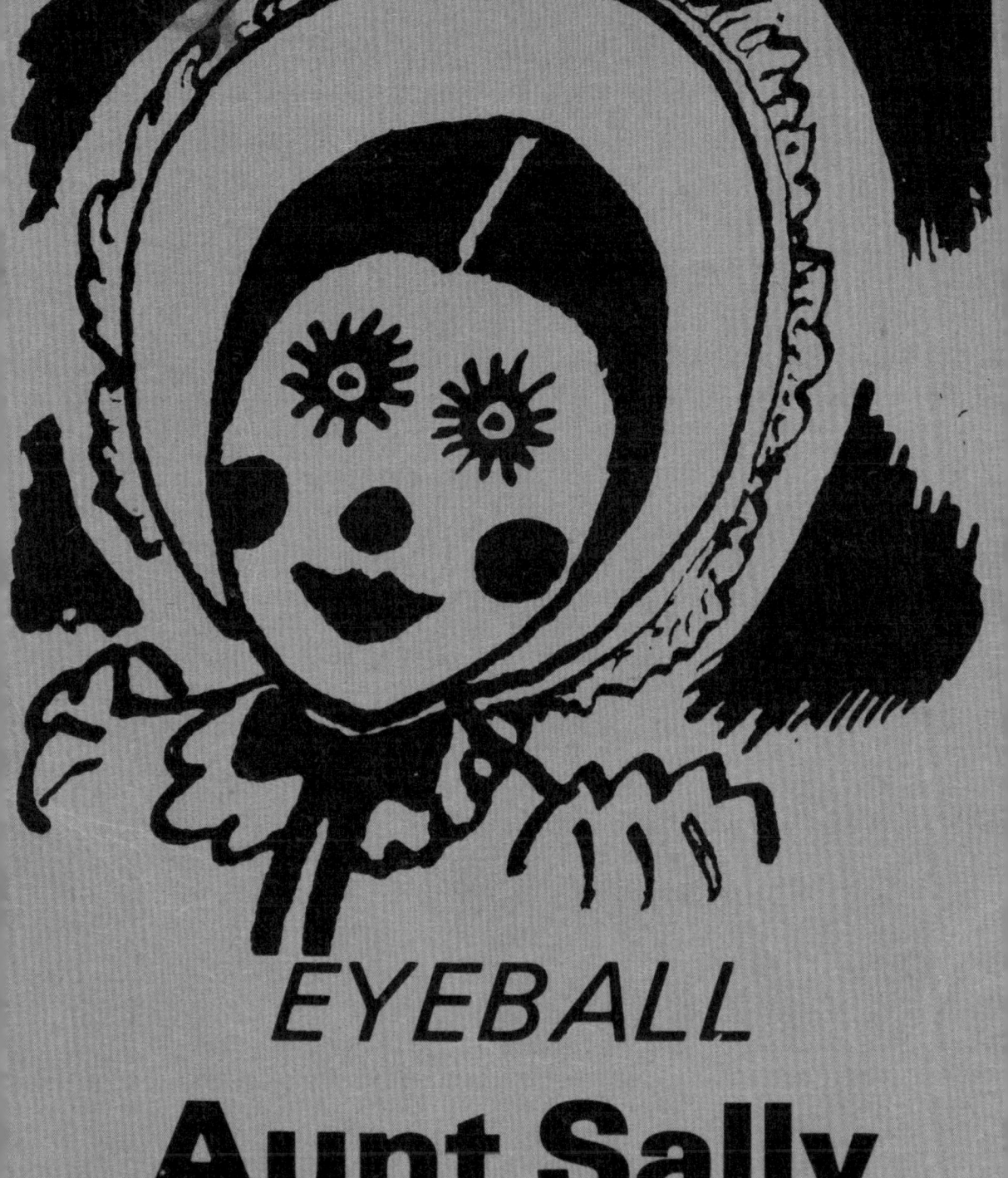

EYEBALL

Aunt Sally

Leiston Breaker

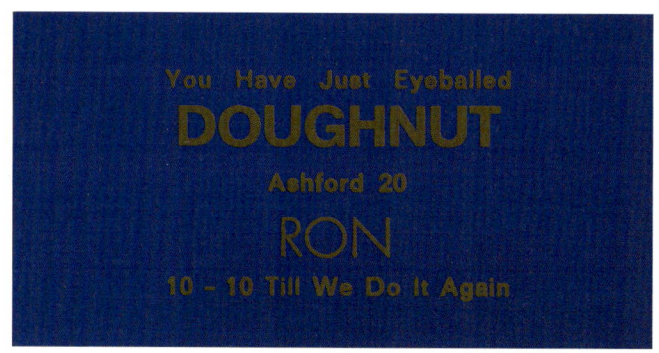

You Have Just Eyeballed

DOUGHNUT

Ashford 20

RON

10 - 10 Till We Do It Again

EYEBALL   EYEBALL
You Have Just Eyeballed

CARROT CRUNCHER

10-10 Till We Meet Again
1 - 40 Be Good Don't Be Naughty

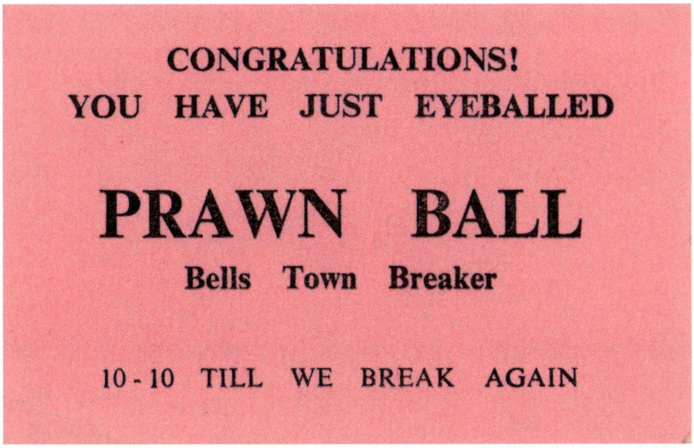

CONGRATULATIONS!
YOU HAVE JUST EYEBALLED

PRAWN BALL

Bells Town Breaker

10 - 10 TILL WE BREAK AGAIN

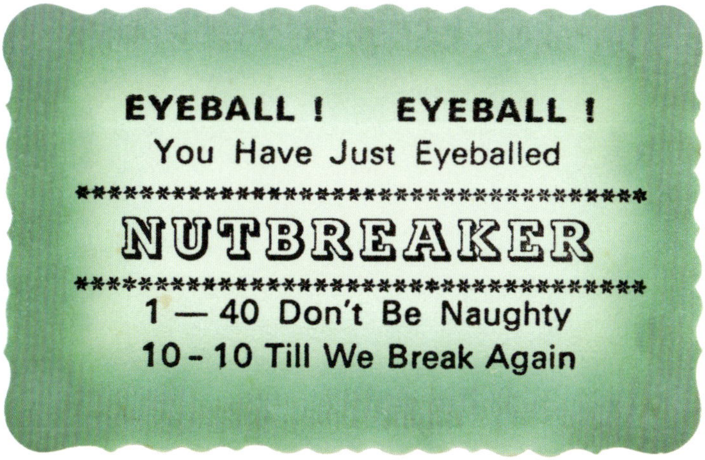

EYEBALL !   EYEBALL !
You Have Just Eyeballed

****

NUTBREAKER

****

1 — 40 Don't Be Naughty
10 - 10 Till We Break Again

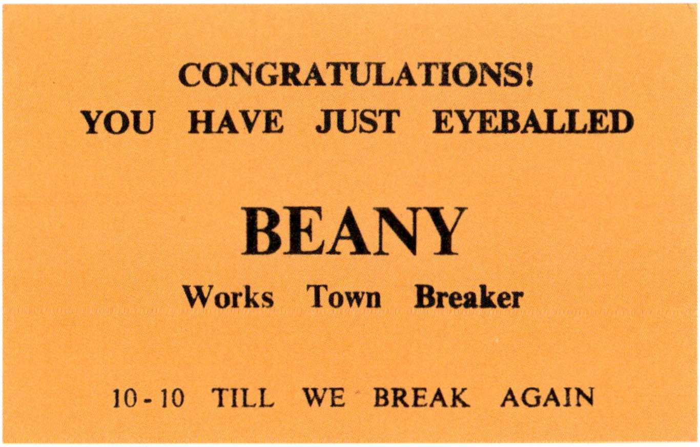

CONGRATULATIONS!
YOU HAVE JUST EYEBALLED

BEANY

Works Town Breaker

10 - 10 TILL WE BREAK AGAIN

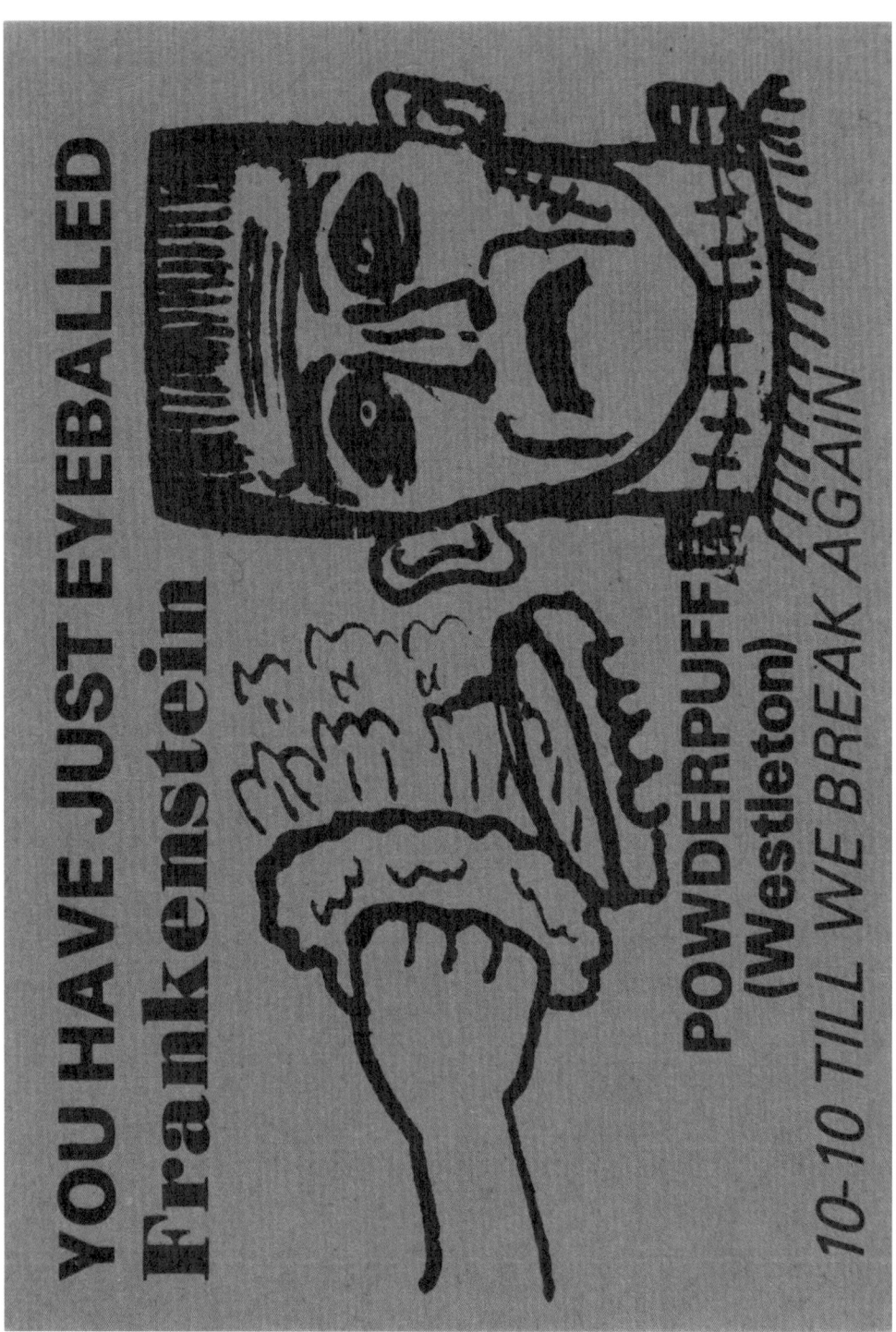

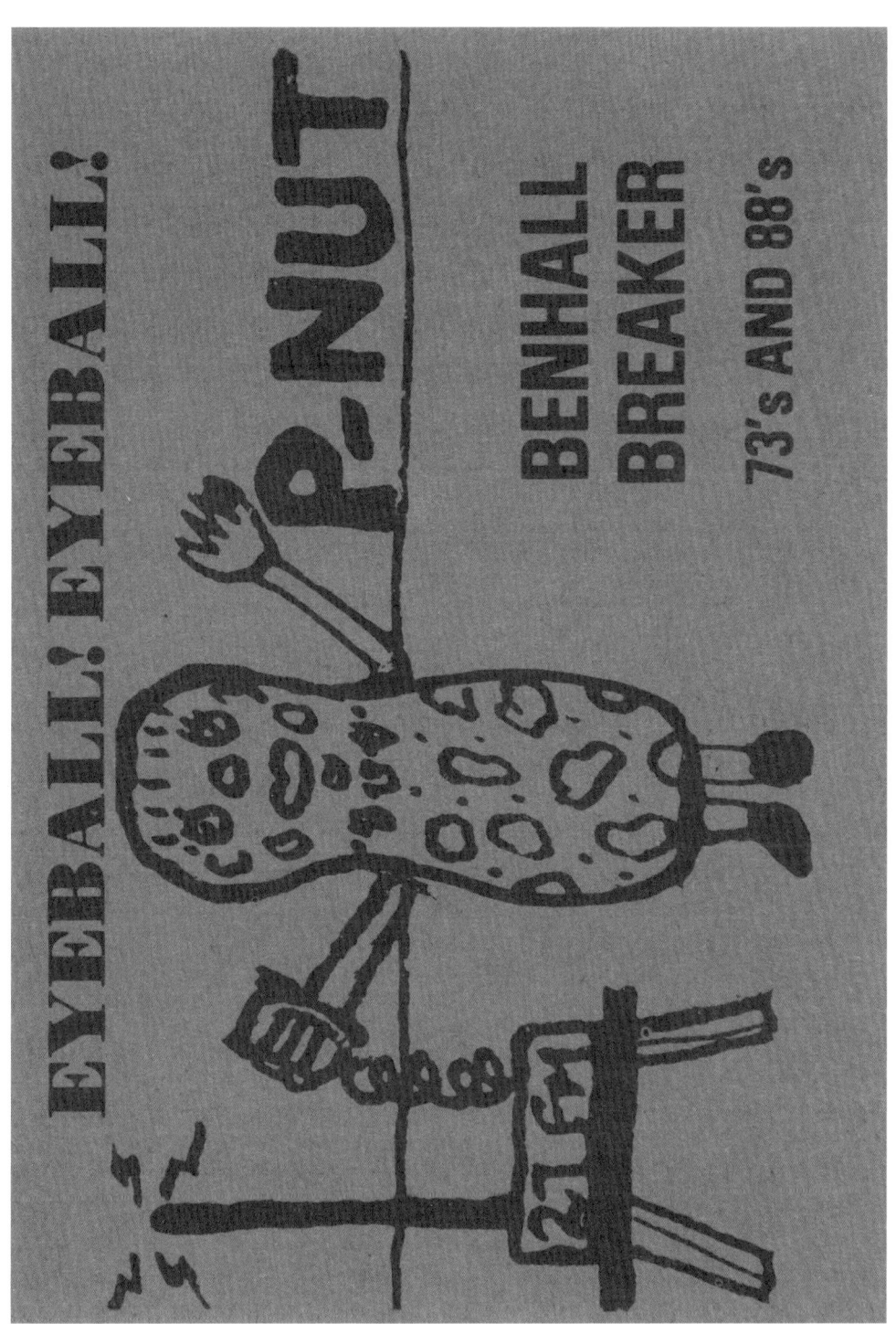

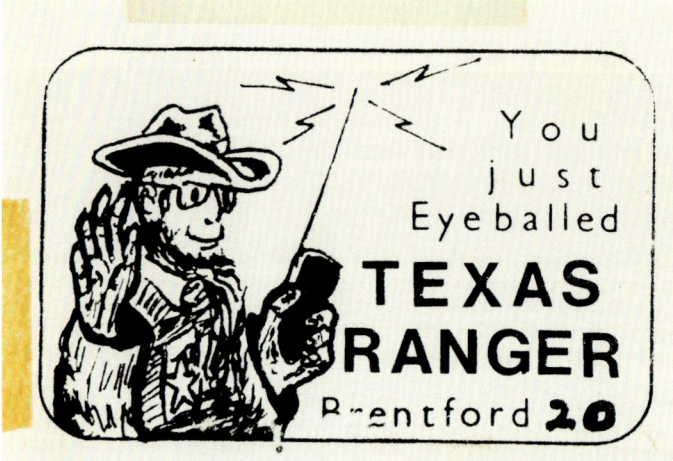

You
just
Eyeballed

TEXAS RANGER

Brentford **20**

EYEBALL! EYEBALL!

You Have Just Eyeballed

SUN SEEKER

Roman City Breaker

10-10 Till We Break Again

CONGRATULATIONS!

YOU HAVE JUST EYEBALLED

KING KONG &
HOUSE MAID

Bruisyard Breakers

10-10 TILL WE BREAK AGAIN

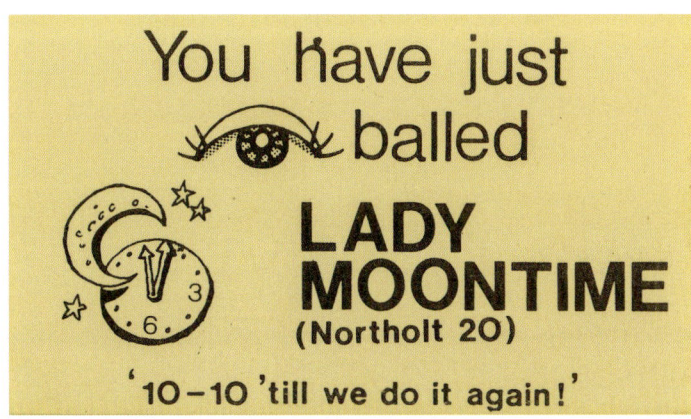

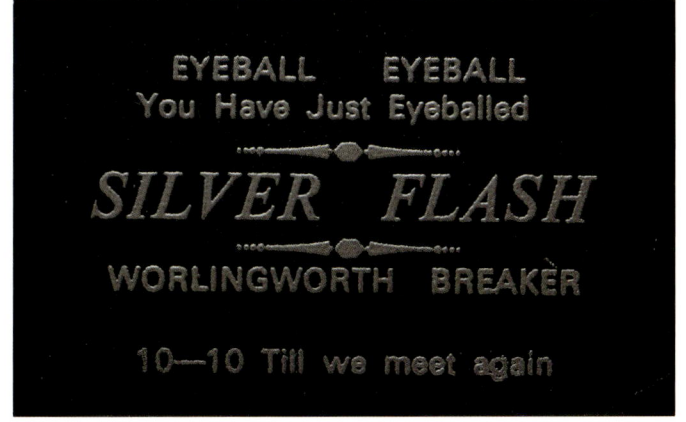

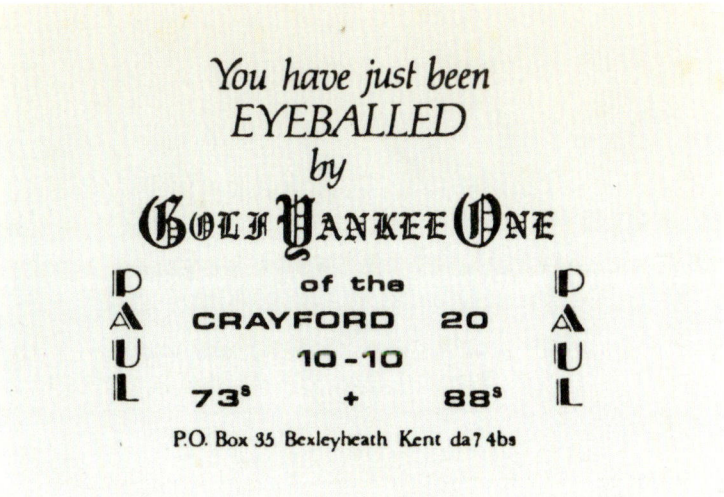

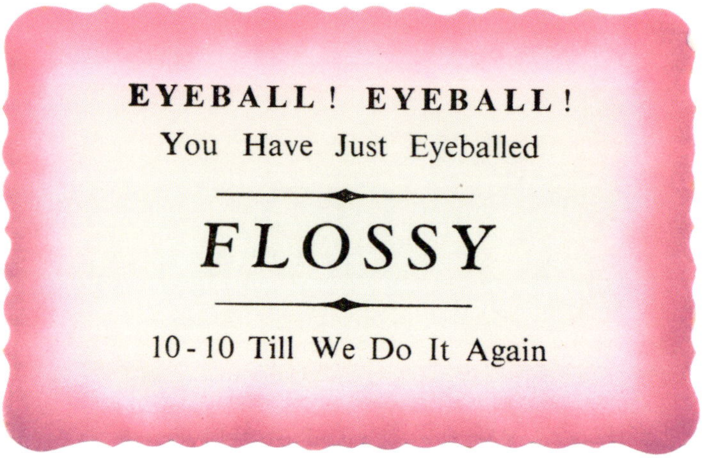

EYEBALL ! EYEBALL !

You Have Just Eyeballed

⬥

FLOSSY

⬥

10 - 10 Till We Do It Again

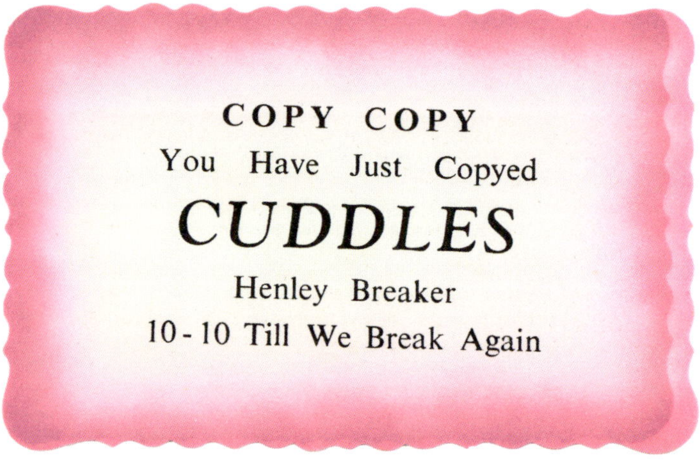

COPY COPY

You Have Just Copied

CUDDLES

Henley Breaker

10 - 10 Till We Break Again

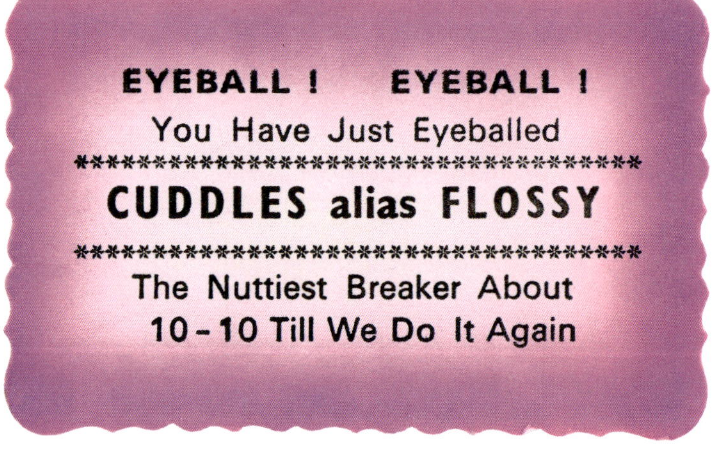

EYEBALL !   EYEBALL !

You Have Just Eyeballed

****

CUDDLES alias FLOSSY

****

The Nuttiest Breaker About

10 - 10 Till We Do It Again

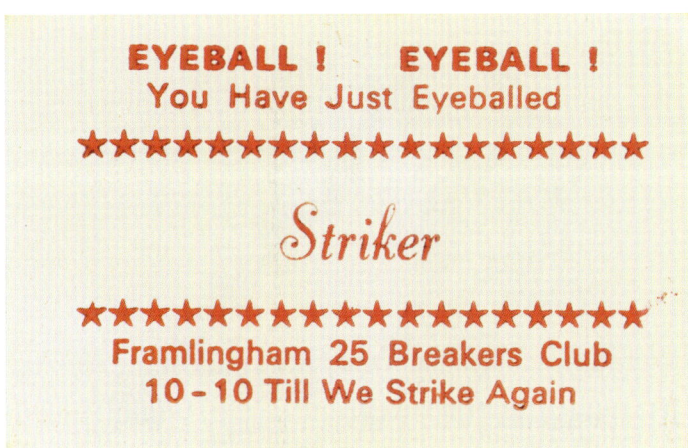

EYEBALL !   EYEBALL !
You Have Just Eyeballed

★★★★★★★★★★★★★★★★★★

Striker

★★★★★★★★★★★★★★★★★★
Framlingham 25 Breakers Club
10 - 10 Till We Strike Again

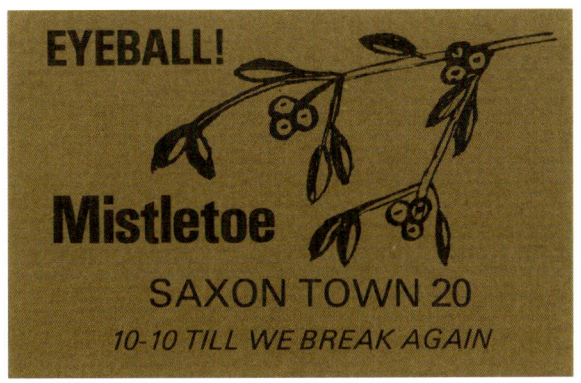

EYEBALL!

Mistletoe

SAXON TOWN 20

10-10 TILL WE BREAK AGAIN

CONGRATULATIONS
You Have Just Eyeballed

WURZEL

Bedingfield Breaker

10 - 10 Till We Do It Again

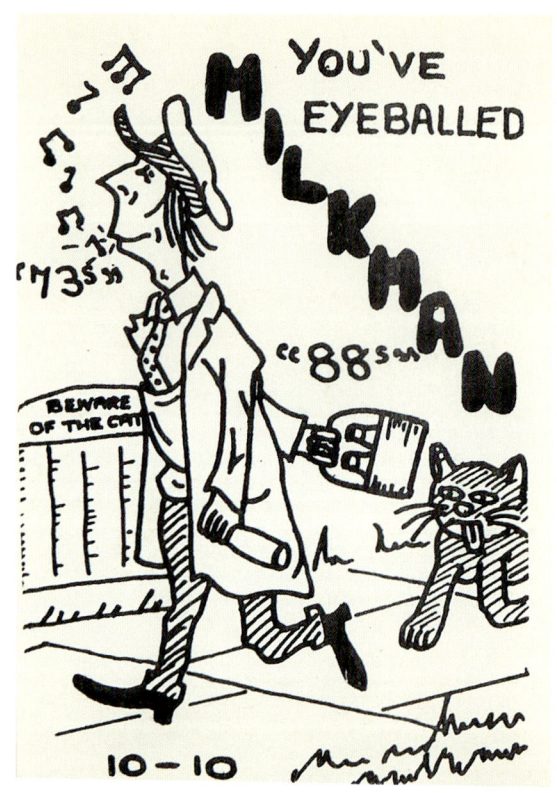

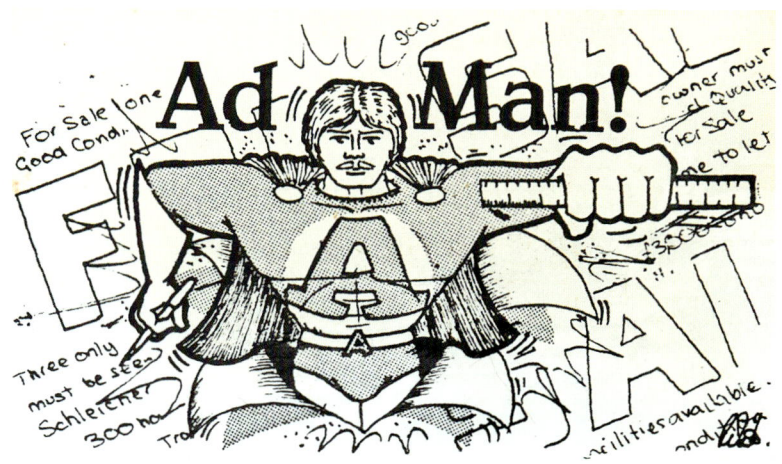

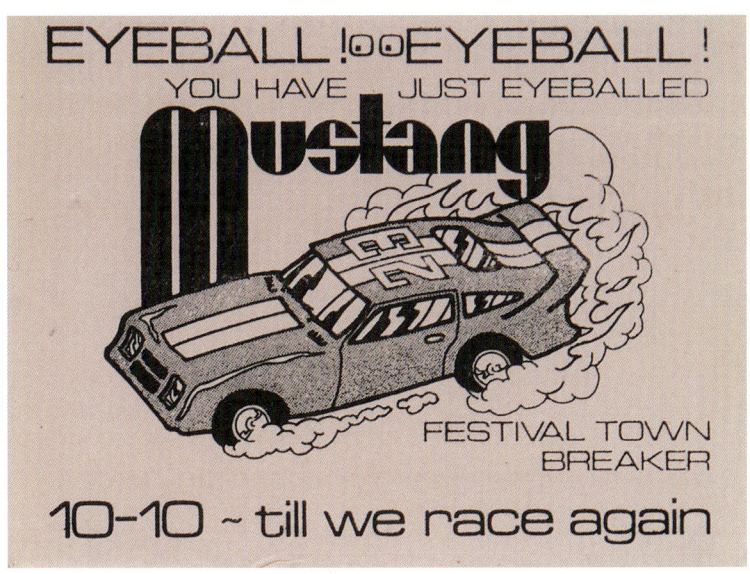

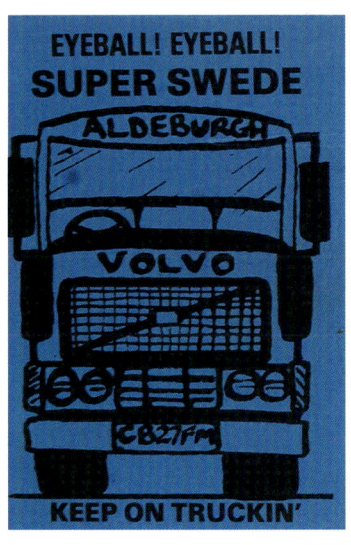

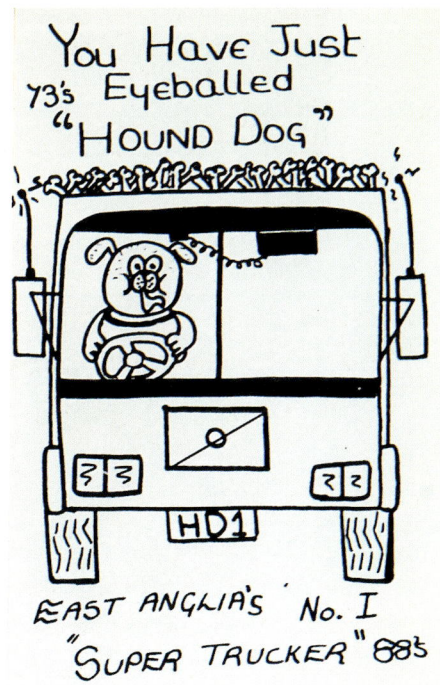

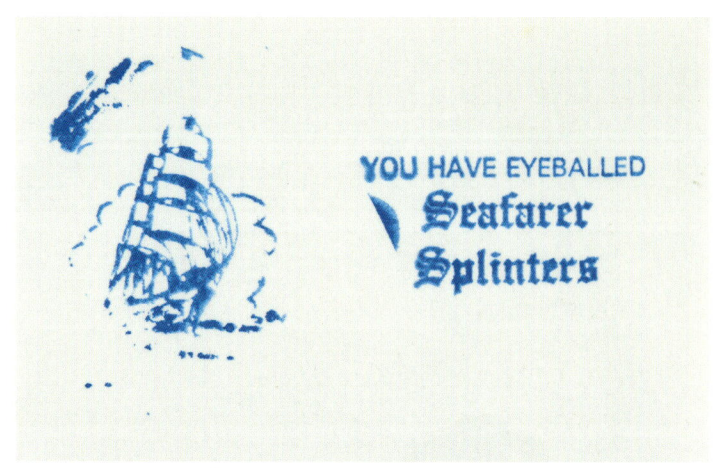

YOU HAVE EYEBALLED
Seafarer
Splinters

EYEBALL! EYEBALL!

You Have Just Eyeballed

SLIPSTREAMER

10-10 Till We Break Again

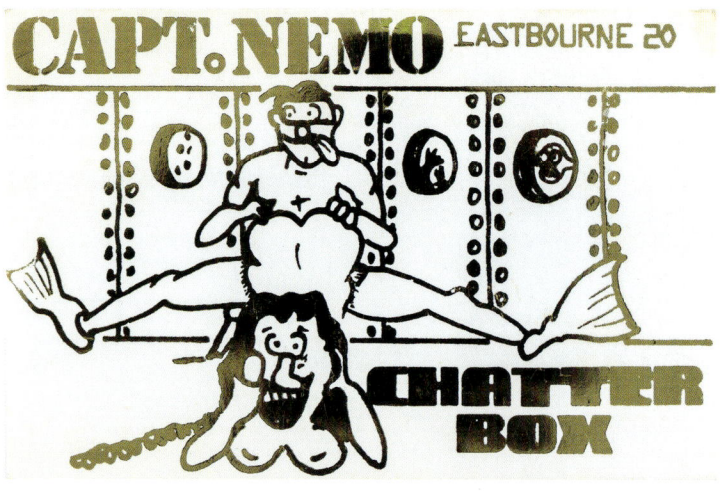

CAPT. NEMO   EASTBOURNE 20

CHATTER BOX

Rubber Duck

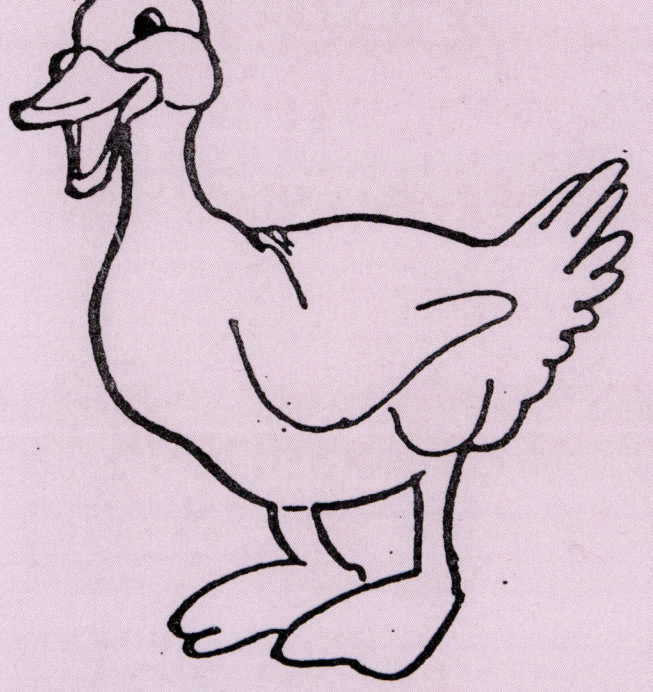

DENNIS

WINDMILL HILL

IT'S YOUR LUCKY DAY YOU HAVE JUST

EYEBALLED

RED BEARD

**UNIT 40
P.O. BOX 5, AMMANFORD, DYFED,
SA18 3BN, UNITED KINGDOM.**

Have a nice day

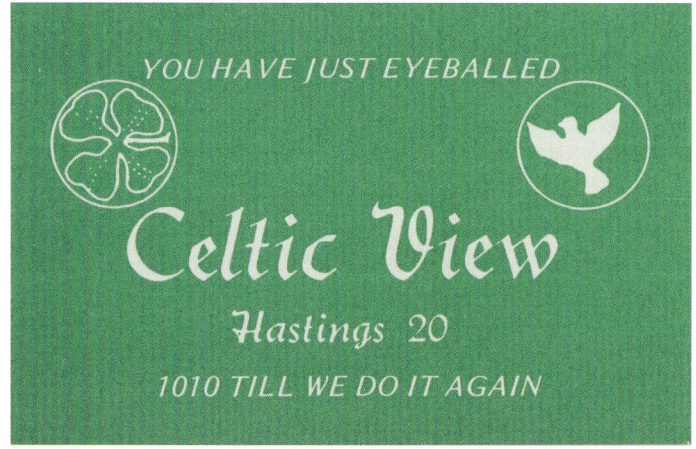

YOU HAVE JUST EYEBALLED

Celtic View

Hastings 20

1010 TILL WE DO IT AGAIN

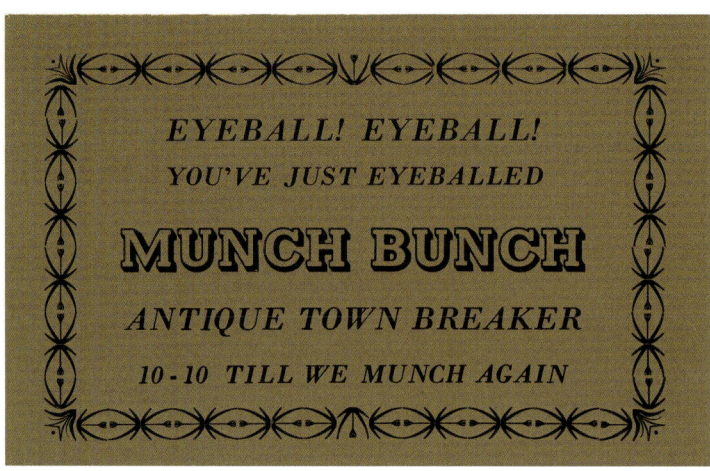

EYEBALL! EYEBALL!
YOU'VE JUST EYEBALLED

MUNCH BUNCH

ANTIQUE TOWN BREAKER

10-10 TILL WE MUNCH AGAIN

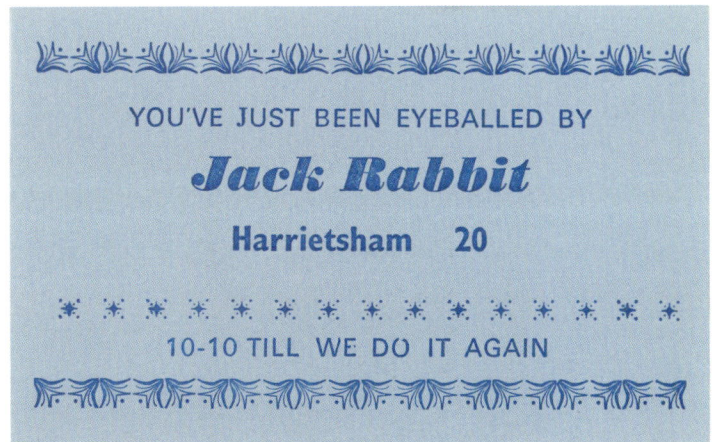

YOU'VE JUST BEEN EYEBALLED BY

Jack Rabbit

Harrietsham   20

10-10 TILL WE DO IT AGAIN

EYEBALL! EYEBALL!

You Have Just Eyeballed

BLUE FOX

Manningtree Breaker

10-10 Till We Break Again

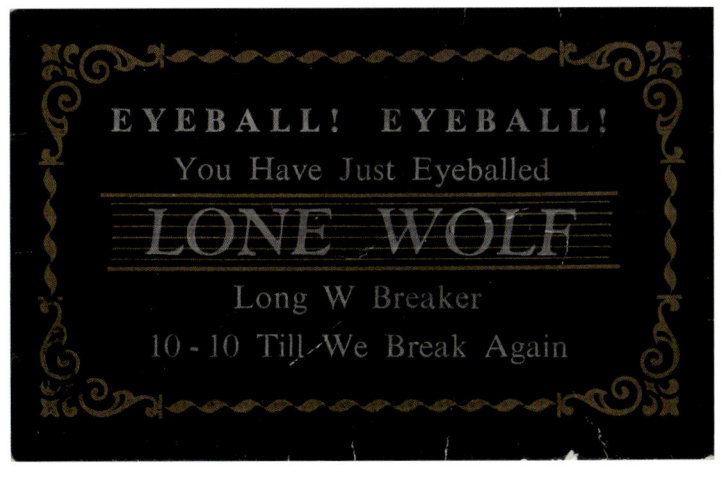

EYEBALL!   EYEBALL!

You Have Just Eyeballed

LONE WOLF

Long W Breaker

10-10 Till We Break Again

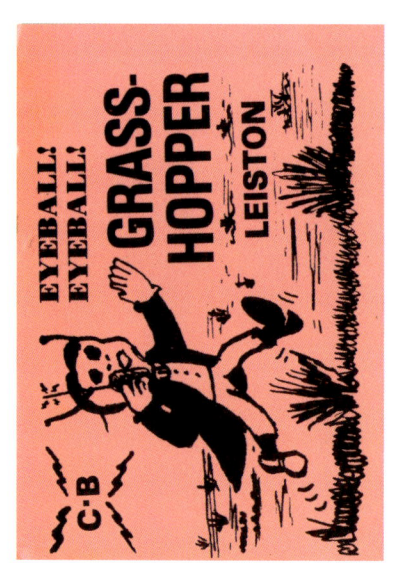

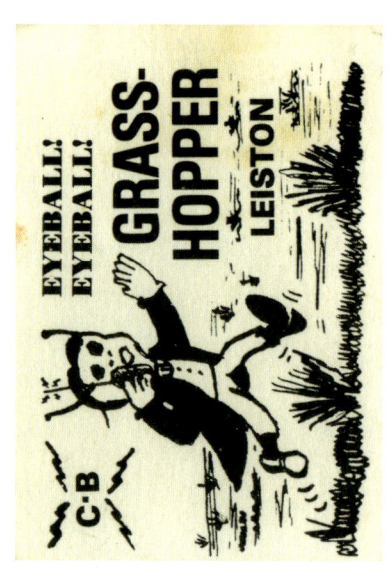

Eric 'Shaver' Cracknell, aka 'Grasshopper'.

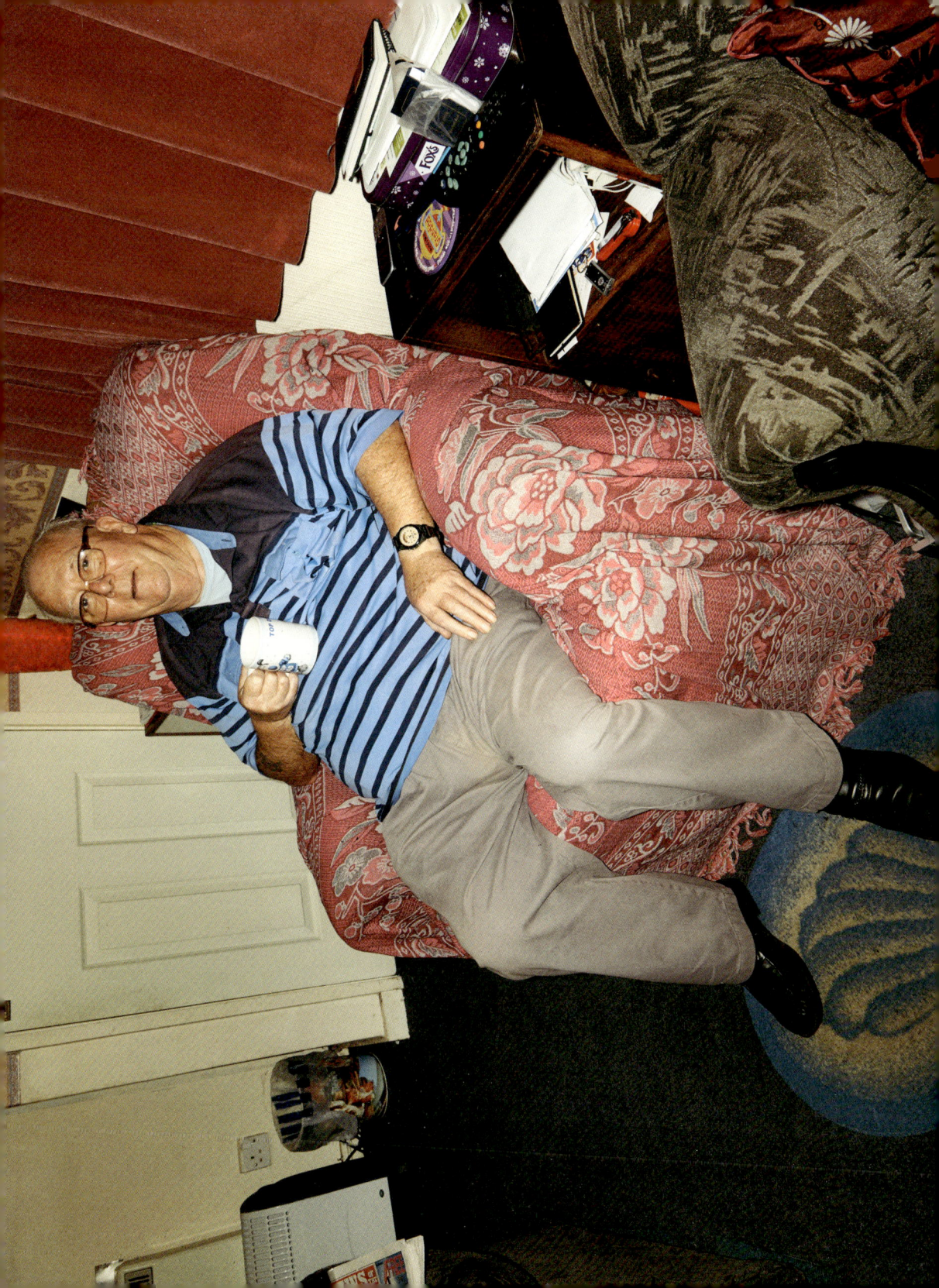

CONGRATULATIONS
You Have Just Eyeballed
THE HULK
Bedingfield Breaker
10 - 10 Till We Do It Again

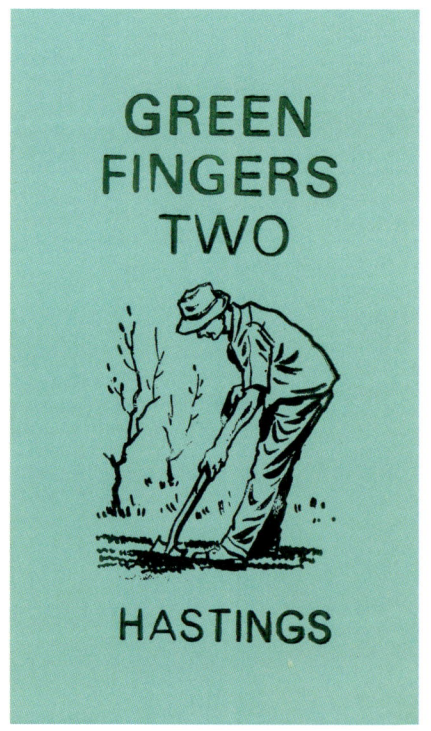

GREEN
FINGERS
TWO

HASTINGS

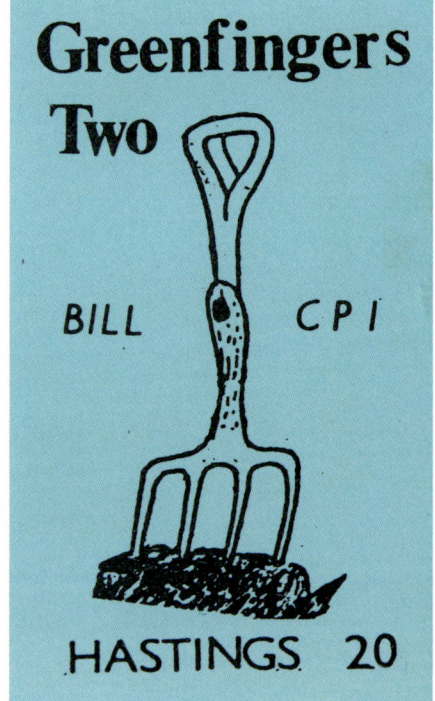

Greenfingers
Two

BILL   CP1

HASTINGS 20

WELL   DONE
You Have Just Eyeballed

THE HULK

BEDINGFIELD   BREAKER

10 - 10 Till we do it again

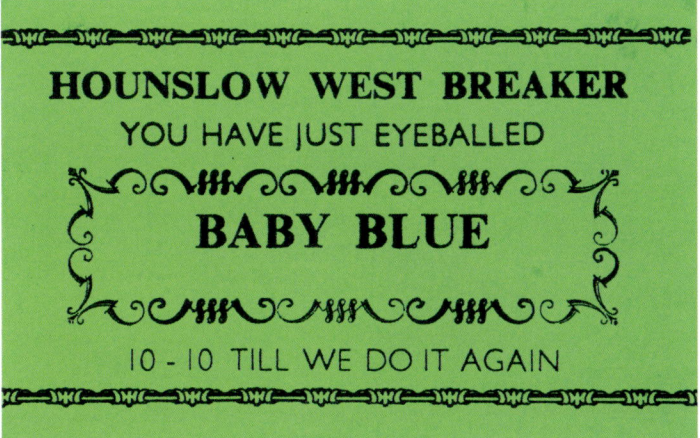

HOUNSLOW WEST BREAKER
YOU HAVE JUST EYEBALLED

BABY BLUE

10 - 10 TILL WE DO IT AGAIN

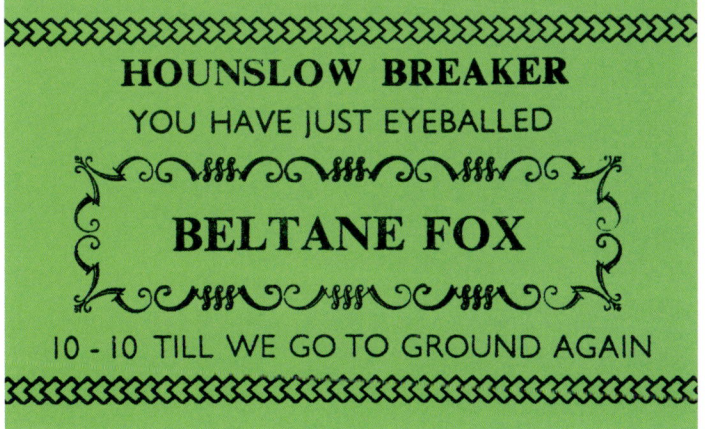

HOUNSLOW BREAKER
YOU HAVE JUST EYEBALLED

BELTANE FOX

10 - 10 TILL WE GO TO GROUND AGAIN

EYEBALL   EYEBALL
You Have Just Eyeballed
GOLDEN SOVEREIGN
&
SLEEPING BEAUTY
LAXFIELD BREAKERS
10-10 Till We Glitter Again

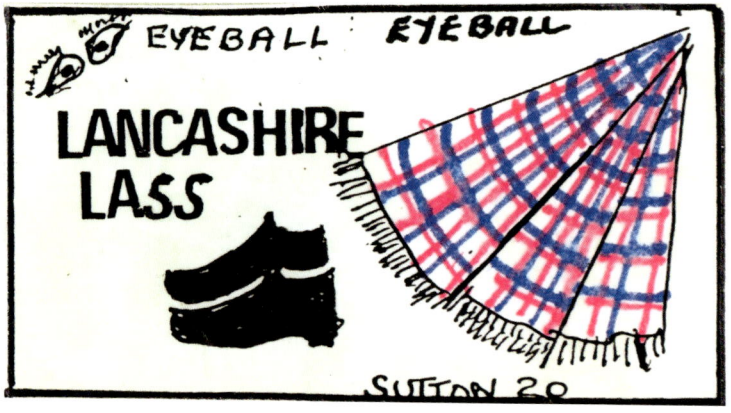

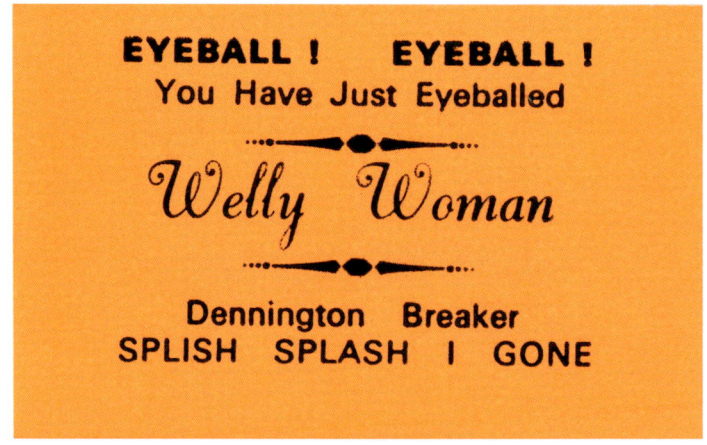

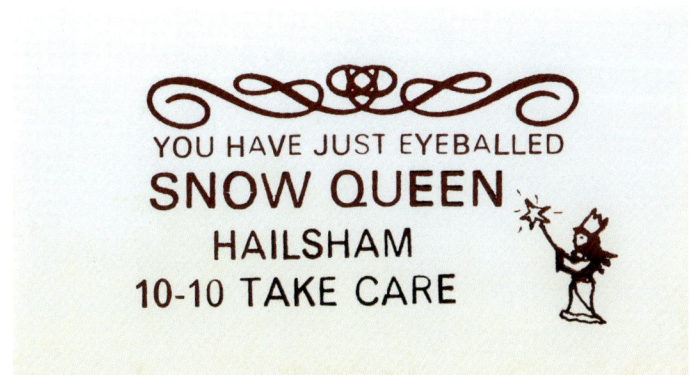

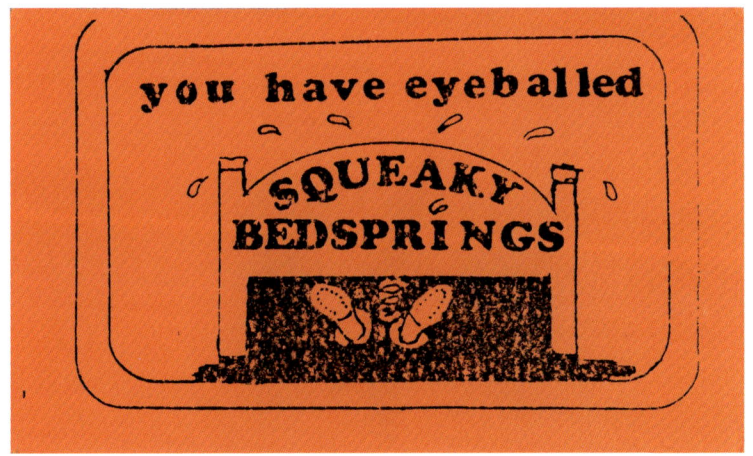

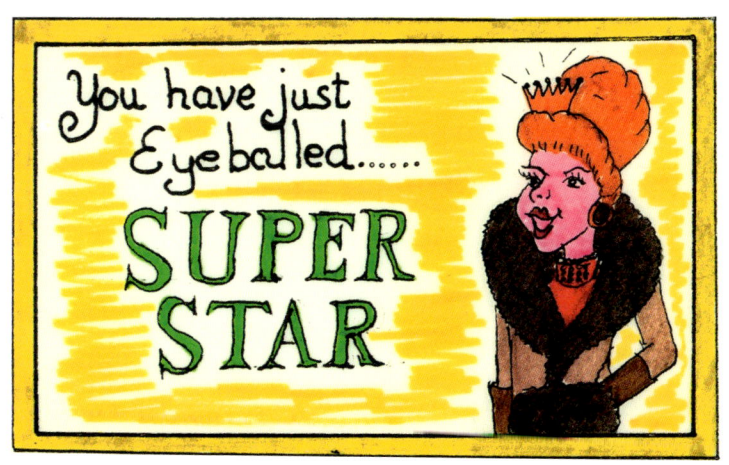

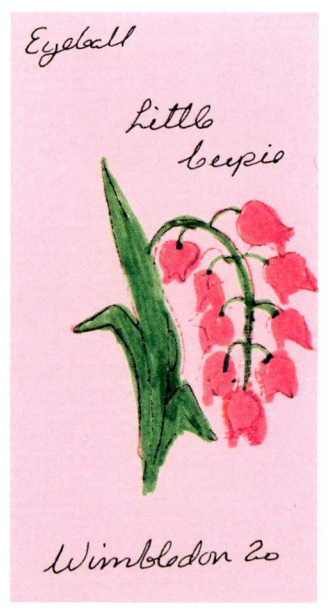

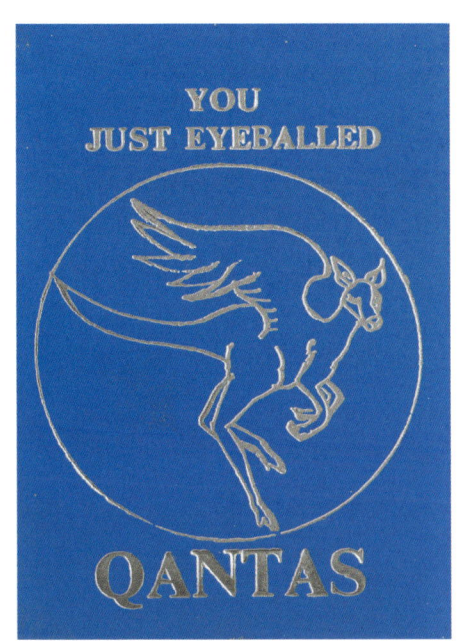

YOU
JUST EYEBALLED

QANTAS

eyeball eyeball

this is

TINY TEARS

sandwich twenty

ON CB YOU CAN DO IT IN ANY POSITION

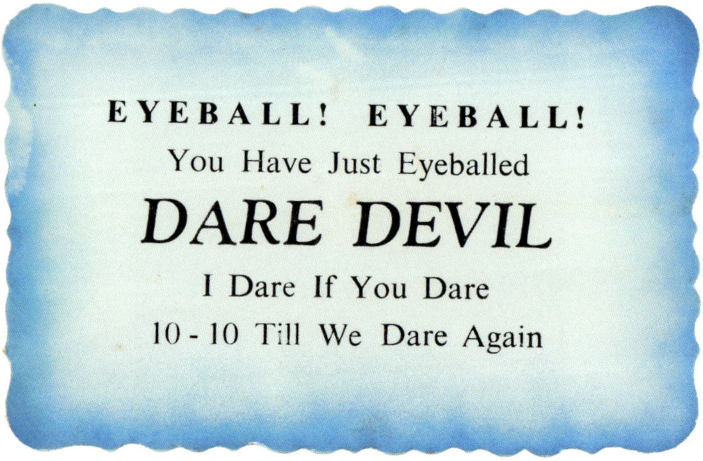

EYEBALL! EYEBALL!

You Have Just Eyeballed

DARE DEVIL

I Dare If You Dare

10 - 10 Till We Dare Again

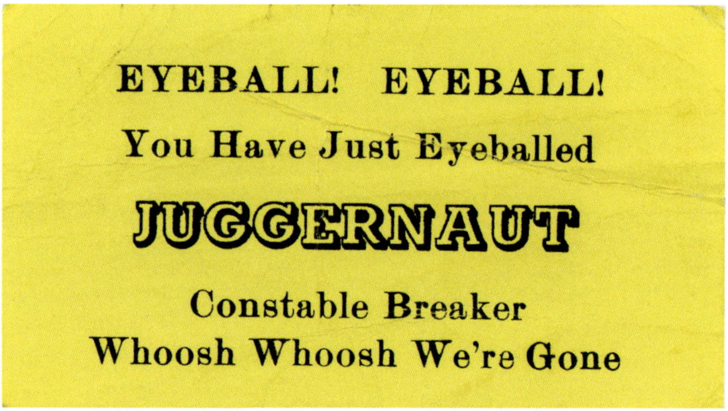

EYEBALL! EYEBALL!

You Have Just Eyeballed

JUGGERNAUT

Constable Breaker

Whoosh Whoosh We're Gone

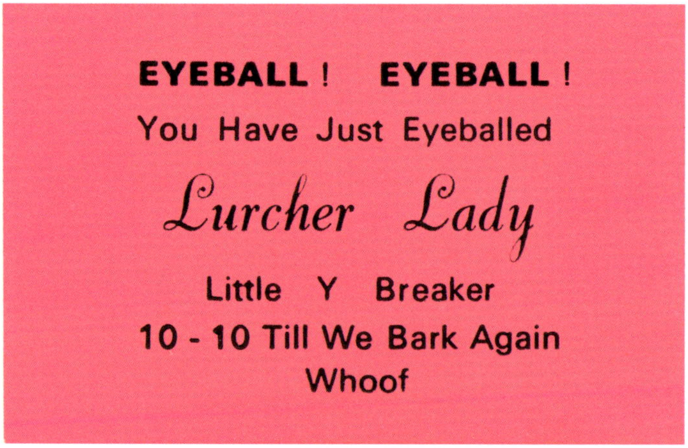

EYEBALL ! EYEBALL !

You Have Just Eyeballed

Lurcher Lady

Little Y Breaker

10 - 10 Till We Bark Again

Whoof

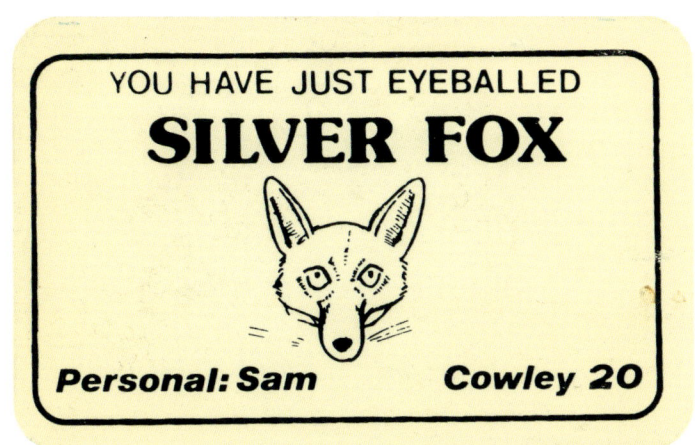

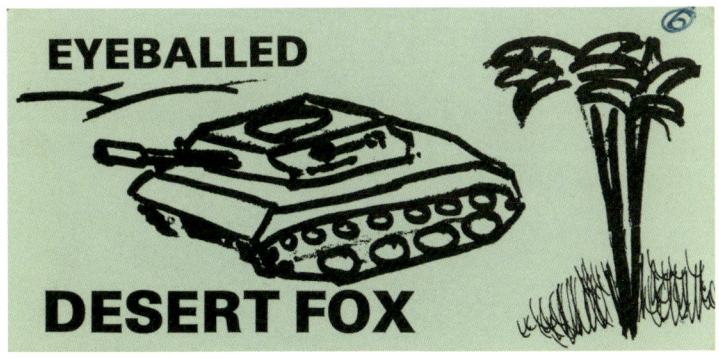

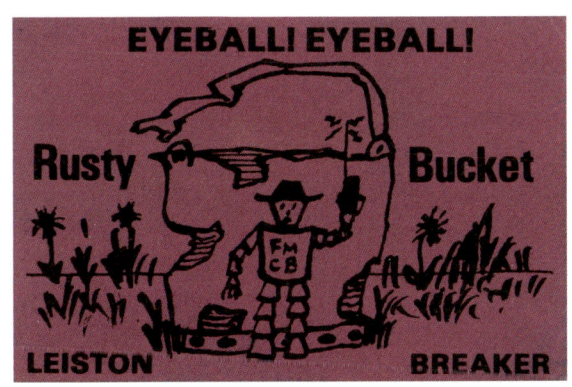

SNOWB

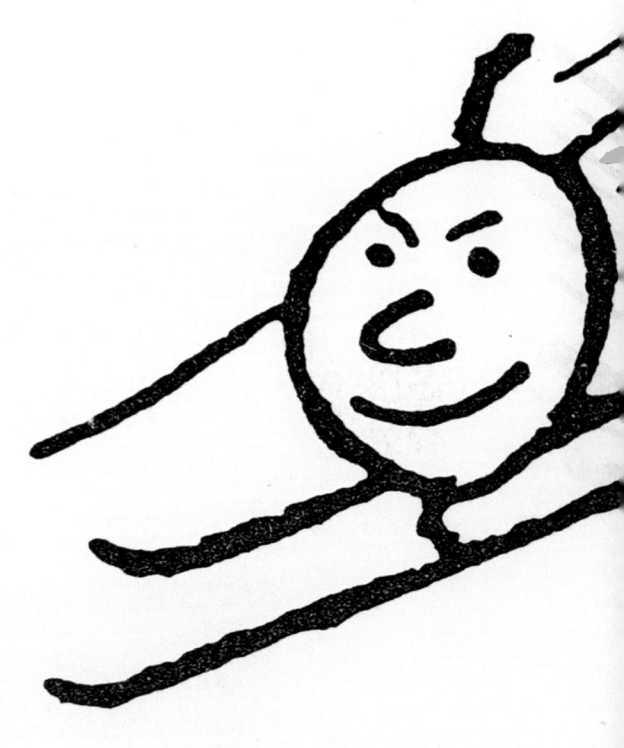

Hast

ERRY

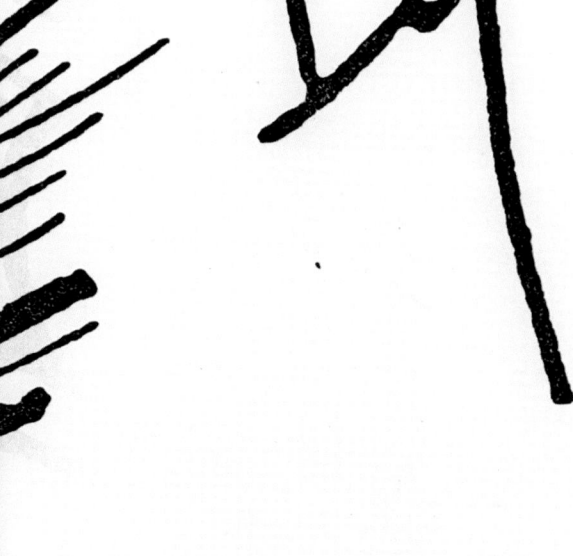

ings

YOU'VE JUST BEEN EYEBALLED BY

Flakey Jake

10-10—TILL WE DO IT AGAIN

» *eyeball ! eyeball* «

YOU HAVE JUST EYEBALLED

SUPER NOVA

10-10 TILL WE BREAK AGAIN

EYEBALL !   EYEBALL !

You Have Just Eyeballed

★★★★★★★★★★
HAYMAKER
★★★★★★★★★★

LAXFIELD BREAKER

10 – 10 Till We HAYMAKE Again

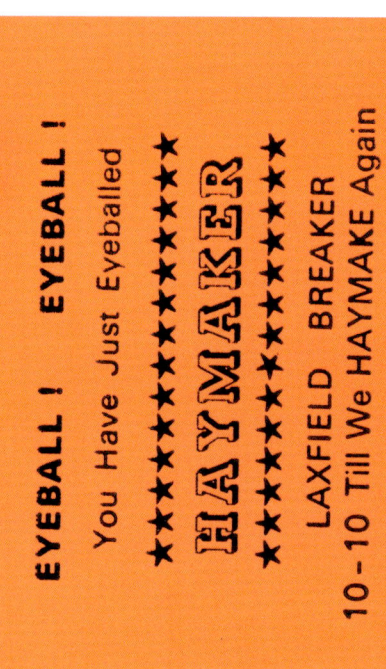

HAIRDRESSER

GOLF CITY

20

10 - 4

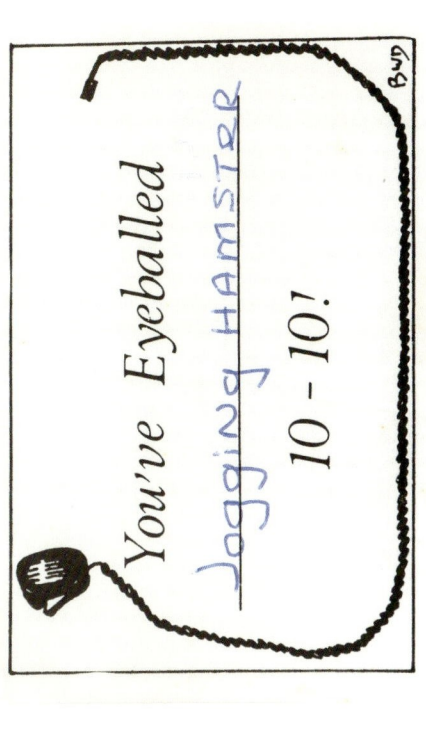

You've Eyebballed

Jogging HAMSTER

10 – 10!

BJP

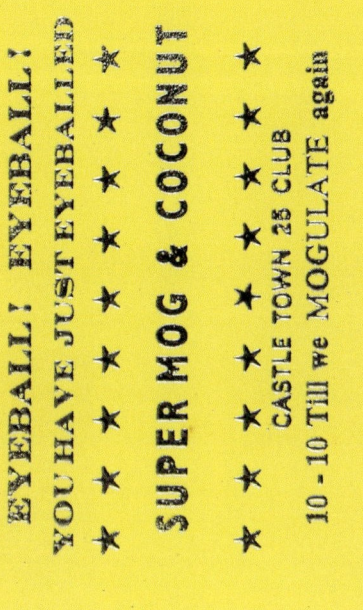

EYEBALL! EYEBALL!
YOU HAVE JUST EYEBALLED

SUPER MOG & COCONUT

CASTLE TOWN 23 CLUB
10 – 10 Till we MOGULATE again

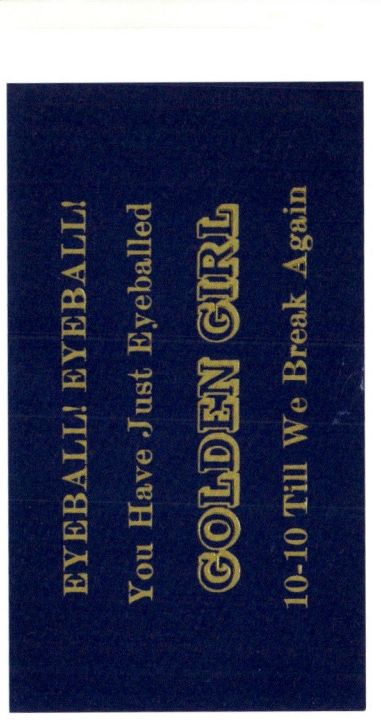

EYEBALL! EYEBALL!
You Have Just Eyeballed

GOLDEN GIRL

10-10 Till We Break Again

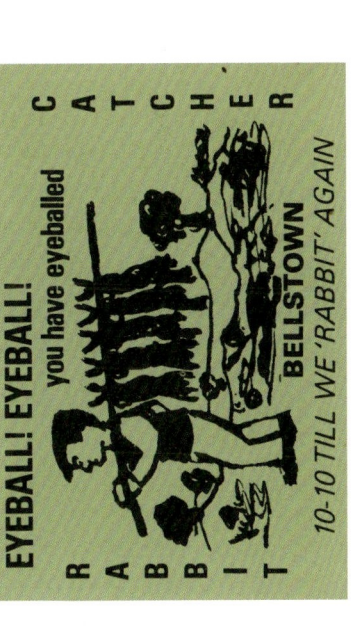

EYEBALL! EYEBALL!
you have eyeballed

CATCHER
RABBIT

BELLSTOWN
10-10 TILL WE 'RABBIT' AGAIN

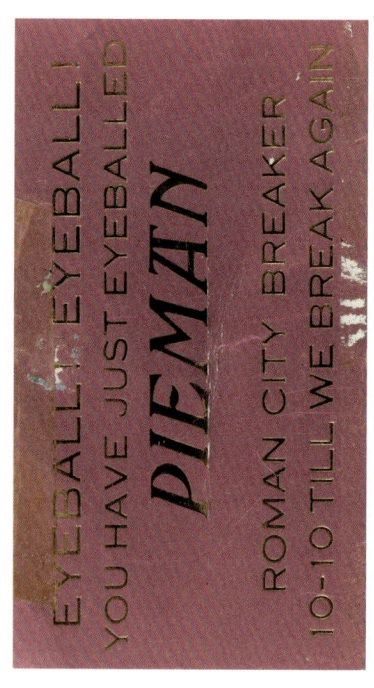

EYEBALL! EYEBALL!
YOU HAVE JUST EYEBALLED
PIEMAN

ROMAN CITY BREAKER
10-10 TILL WE BREAK AGAIN

Simon Morgan, aka 'Pieman'.

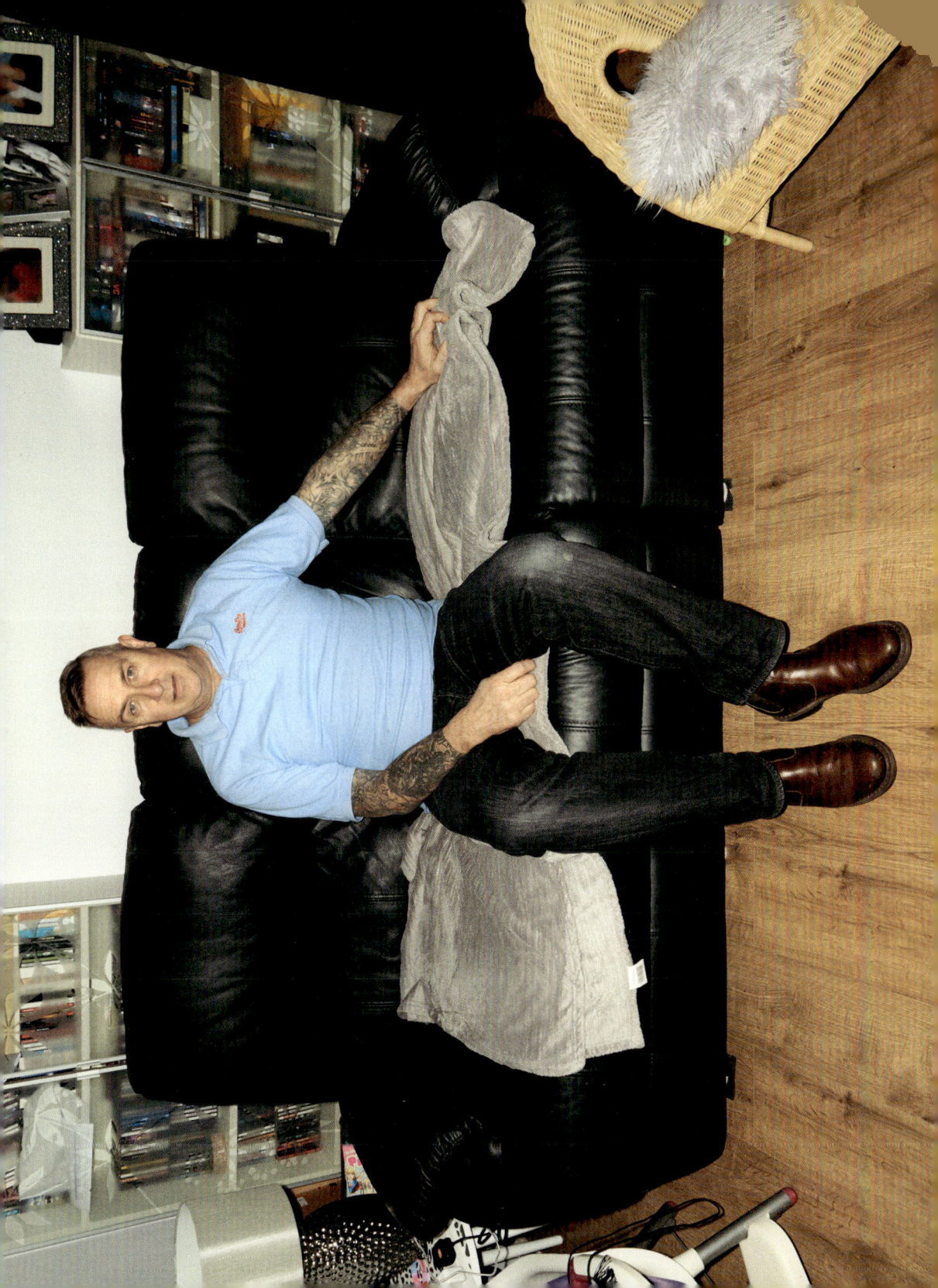

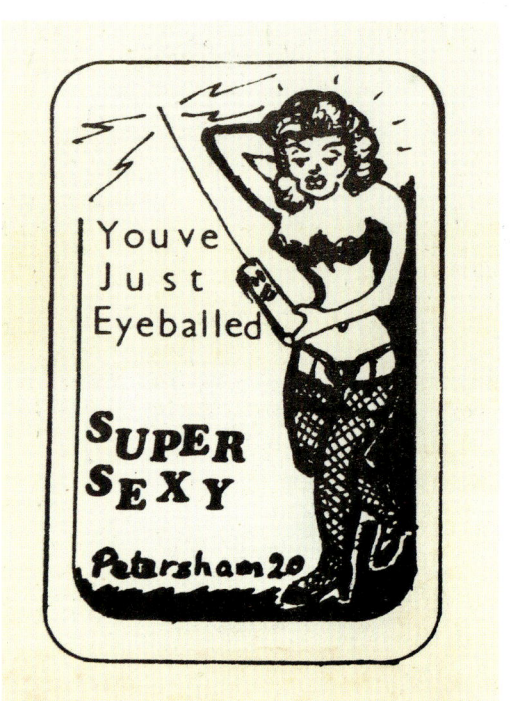

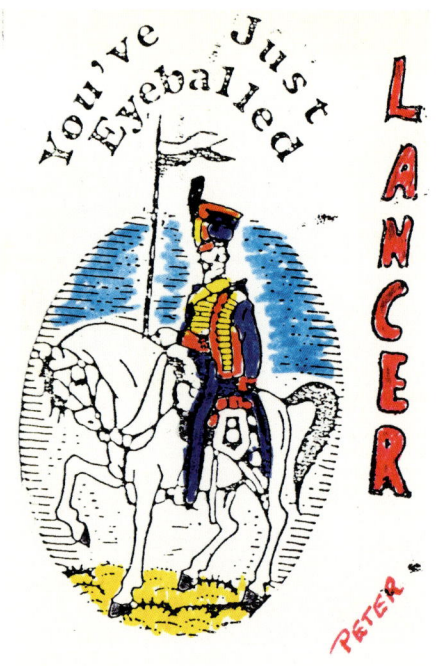

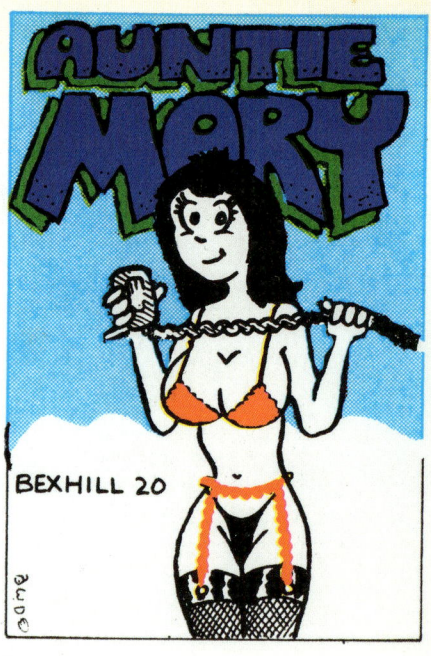

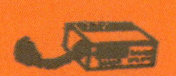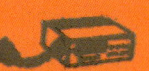

YOU JUST EYEBALLED

Welsh Dragon
& Nightingale
HAILSHAM 20
10 10 TAKE CARE

EYEBALL! EYEBALL!

You Have Just Eyeballed

TRAILBOSS &
GREEN EYES

Ipswich (Big I) Breaker
10-10 Till We Break Again

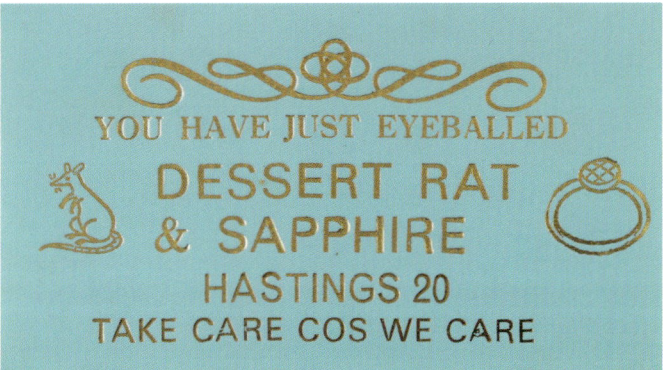

YOU HAVE JUST EYEBALLED

DESSERT RAT
& SAPPHIRE
HASTINGS 20
TAKE CARE COS WE CARE

You ha

eyeb

Black

ar

Cre

ue just
alled
Coffee
nd
eam

Peanut butter in your ears

Irreverent, coded, and rich with playful imagery, the phrases of CB radio make up a pidgin language all of its own. When in full flow, it can include a heady mix of amateur radio's Q code and elements of the police 10 code, but usually it's a scattergun of accepted half-sense and carriageway wisdom.

Alligator	Someone who doesn't reply
Bubble trouble	Tyre problems
Bucket mouth	Person who swears on air
Down and on the side	Finished talking but still listening
Everybody's walking the dog	Channels are busy
Get your ears on	Listen in
Handle	Name
Idiot box/One-eyed monster	TV
Jaw jacking	Talking profusely
Keep your nose between the ditches and Smokey out of your britches	Travel safely
Kojak with a Kodak	Policeman operating radar
Lettuce	Money
Mama Bear	Female police officer
Motion lotion	Fuel
On the side	Asking to break into a conversation
Over your shoulder	Behind you
Panic on the streets	An area monitored by the Post Office
Pavement princess	Prostitute
Peanut butter in your ears	Not listening to CB radio
Permanently 10–7	Dead
Portable farmyard	Cattle truck
Pregnant rollerskate	VW Beetle
Quick trip around the horn	Scan channels for activity
Reading the mail	Listening in
Riot squad	Neighbours experiencing TV interference
Shaking the windows	Loud and clear
Shoes/Socks	Linear amplifier

Six lane car park/Superslab	Motorway
Smokey	Police
Snafu	Cock-up
Split your sides	Transmit on Sideband
Tighten up the rubber band	Accelerate
Tin can	CB rig
Twig	Aerial
Wall to wall spaghetti	Skip/crosstalk from Italy
Wall to wall and tree top tall	Loud and clear
10-4	Affirmative
10-6	Busy
10-8	In service/ready
10-9	Repeat requested
10-10	Finished broadcasting but will continue to listen/take care
10-33	Emergency situation
10-40	Please tune to channel...
10-51	Headed your way
20	Location/as in 'what's your twenty?'
73s	best wishes/love
88s	kisses

Towns and cities earned their own handles through large local industries, and sometimes just playful associations. It's another imported and converted part of US CB culture that found its written form on Eyeball Cards to show their '20' (location). Below is a list of some of the places on cards featured in this book, as well as other major towns and cities.

Bay Town	Morecambe
Big C	Croydon
Big G	Gloucester
Black Pudding City	Bury
Booze City	Portsmouth
Canary Town	Norwich
Castle City	Edinburgh
Castle Town	Skipton
Cathedral City	Lincoln
Chicken Town	Bradford on Avon/Eye
Chip Butty Town	Bolton
Cider City	Hereford
Cider Town	Taunton
Cotton Town	Preston
Customville	Sleaford
Dead City	Birmingham
Dreamy Town	Luton
Eastside	Barking
Festival Town	Aldeburgh
Fine City	Norwich
Foggy Town	London
Ghost Town	Corby/St Ives/Swindon
Golf City	St Andrews
Golden Mile	Great Yarmouth
Granite City	Aberdeen
Lillipda	Isle of Sheppey
Milltown	Oldham
Murky River City	Liverpool
New Town	Skelmersdale
Night City	Horley
Noddy Town	London

Quarry Town	Heywood
Rail Town	Crewe
Rainy City	Manchester
Roman City	Bath
Salt City	Winsford
Second City	Glasgow
Smelly Town	Bridgewater
Smoke City/Smokey Town	London
Smokey Dragon	Cardiff
Soul City	Belfast
Spire City	Salisbury
Steel City	Sheffield
Sugar Town	Bury St Edmunds
Sunny Bay City	Torquay
Surf City	Portrush
Works Town	Leiston
Apple County	Somerset
Bridge County	Humberside
Cream County	Devon
Hop County	Kent
Royal County	Buckinghamshire
Shoe County	Northamptonshire
Surf County	Cornwall

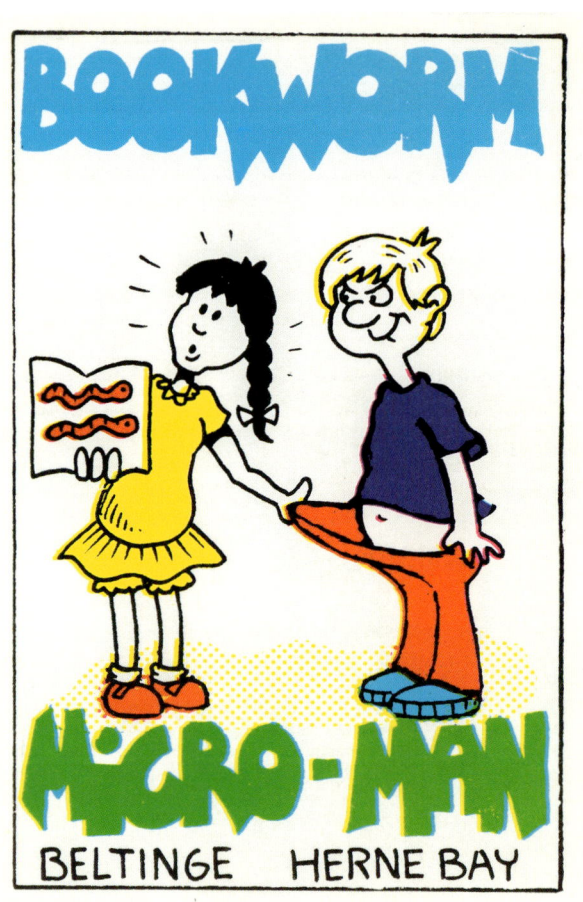

Acknowledgements and thanks

Thanks to Alan Watt, Jordan Smith, Justin Quirk, Rob Fearne, UK CB & Amateur Radio Facebook page, and Mark Atkinson.

Special thanks go to the breakers: Simon Morgan 'Pieman', Jenny Dye 'Fruitcake', Bob Dye 'Sugarbeet', Martin Cook 'Bandit', Mark Alden 'Praying Mantis', Mark Cracknell 'Midget Man', Eric 'Shaver' Cracknell 'Grasshopper', Peter Hurren 'Dart Player', Denise Cracknell, and the unnamed CB-ers who wish to be known only as 'Black Hat', 'Cowboy', 'Dandelion', 'Fidget', 'Redneck' and 'Hummingbird'.

Eyeball Card contributors: David Titlow, Jonny Trunk, Andy Newham, Denise Cracknell, Jordan Smith from Cardboard America, and Simon Morgan.

We'd like to express special thanks to Jonny Trunk for sharing his collection of cards and showing his support for the project right from the word 'breaker'.

The authors would also like to share all the good numbers with Elinor Jansz and Richard Embray at Four Corners Books, John Morgan and his team, as well as Alan Watt and Sipke Visser. Portrait retouching by Frisian, image reproduction by Martin Chapman.

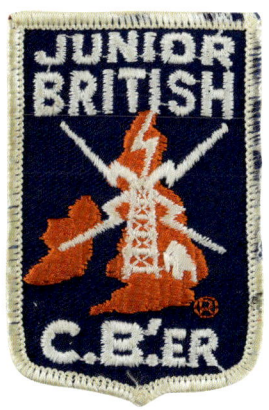

Four Corners Irregulars
A series of books about modern British
visual culture. This is book 1.

Also available:
2. UFO Drawings From
   The National Archives
3. Poster Workshop 1968–1971
...with further volumes in preparation.

Set in Starling and printed
on Garda Ultramatt 130gsm

Published in 2017 by Four Corners Books
56 Artillery Lane, London E1 7LS

Designed by John Morgan Studio
morganstudio.co.uk

Print production by Martin Lee
Reprography by Martin Chapman
Printed in Italy by Printer Trento
Second printing, 2017

Distributed in the UK by Art Data
artdata.co.uk

ISBN 978-1-909829-08-4

10-10 'til we see you again
at fourcornersbooks.co.uk

Page 10, image courtesy Cardboard
America cardboardamerica.org
Page 17, image courtesy Mirrorpix

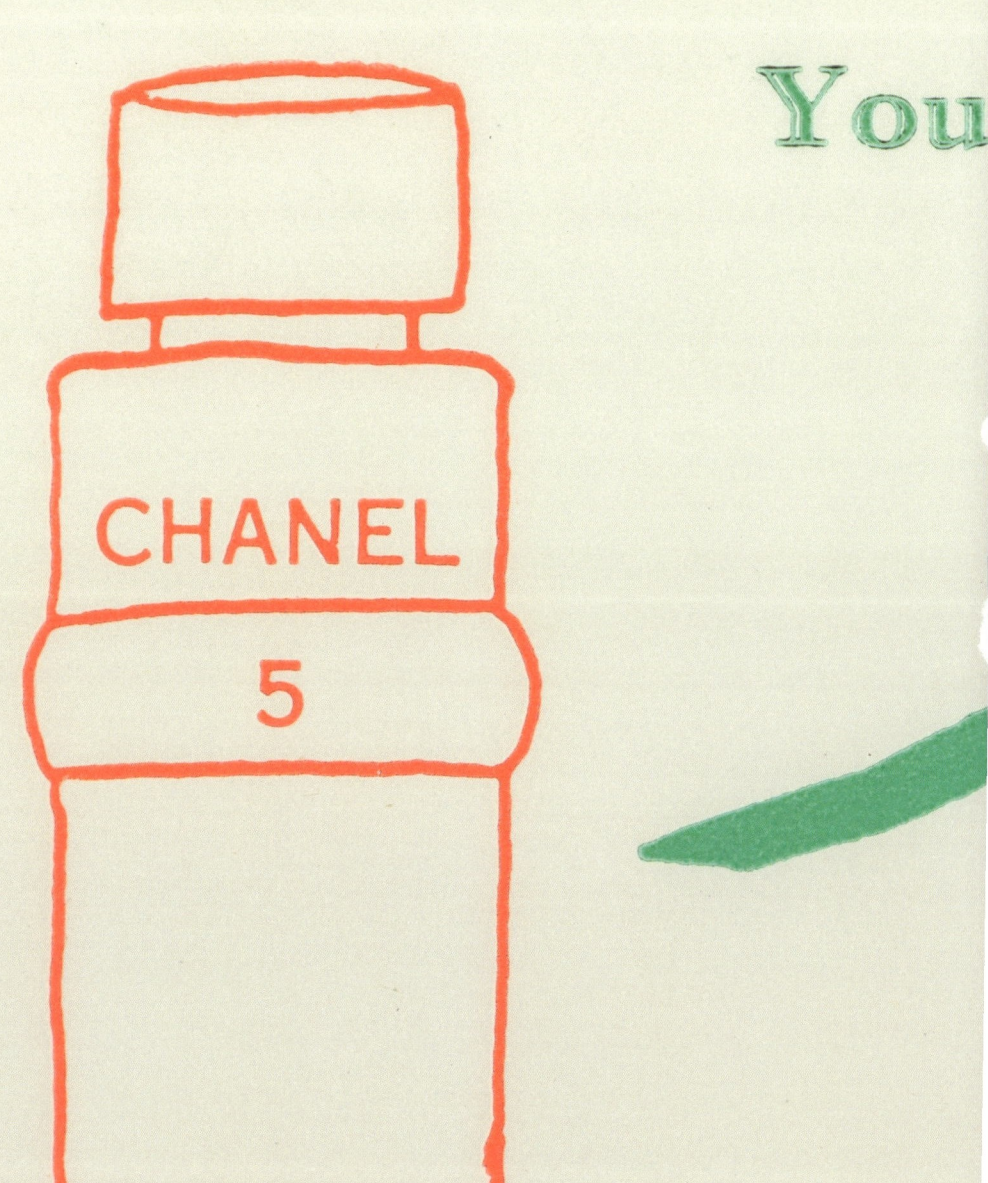

You